# APPROPRIATE
# TECHNOLOGY

· · · · · · · · · · · · · · · · · · ·

*Tools, Choices,*
*and Implications*

# APPROPRIATE
# TECHNOLOGY

· · · · · · · · · · · · · · · · · ·

*Tools, Choices,
and Implications*

## Barrett Hazeltine • Christopher Bull

*Division of Engineering
Brown University
Providence, Rhode Island, USA*

**ACADEMIC PRESS**

San Diego  London  Boston  New York  Sydney  Tokyo  Toronto

Copyright © 1999 by Academic Press

ACADEMIC PRESS
525 B Street, Suite 1900, San Diego, CA 92101-4495, USA
http://www.apnet.com

Academic Press
24–28 Oval Road, London NW1 7DX, UK
http://www.hbuk.co.uk/ap/

**Library of Congress Cataloging-in-Publication Data**

Hazeltine, Barrett.
　　Appropriate Technology : tools, choices, and implications / Barrett
　　　Hazeltine, Christopher Bull.
　　p. cm.
　　Includes bibliographical references and index.
　　ISBN 0-12-335190-1
1. Appropriate technology. I. Bull, Christopher. II. Title.
T49.5.H4 1999
338.9'27–dc21                                                    98-22150
                                                                        CIP

Printed in the United States of America
98 99 00 01 02  IP  9 8 7 6 5 4 3 2 1

• • • • • • • • • •

*This book is dedicated to our families.*

# CONTENTS

# ACKNOWLEDGMENTS

The authors would like to recognize the special contribution of the following individuals, each of whom wrote significant parts of different chapters:

Javed M. Ahmed

Ricardo Bayon

Peter M. Bridge

Float Auma Kidha

Amitabh Pandey

Eric Rudder

Christy L. Stickney

# PREFACE

This book grew out of a course we have been teaching at Brown University since 1980. The course is one of a group of engineering courses for liberal arts students. These courses are not designed to prepare students for further study in engineering. They do not have math or science prerequisites. Rather, using technology as a vehicle, they foster the goals of a liberal arts education — increasing students' analytical ability, their understanding of the contemporary world, their confidence in dealing with complex problems, and their ethical and aesthetic sensitivity. There were several motivations for creating these courses and they continue to motivate our teaching.

Justification for this program of courses was the pervasive influence of technology in our society — it is not possible to understand our society without some understanding of technology. A related need is for a citizenry informed about technology, which is perhaps the most important determinant of national policy. A third justification was a need to make technology accessible to students, so each could relate it to her or his own interests. The course matched an aim of the Alfred P. Sloan Foundation's "New Liberal Arts Program," instituted at about the same time. The Sloan Foundation supported the writing of this book.

Basically the book consists of two parts: technology and implications, or hardware and software. Hardware issues are how to build something, what it costs, and what it produces. Software issues are what people must do to make the hardware effective and, by implication, what can happen if needs of people are not considered. We will see that full attention must be paid to both hardware and software if effective change is to occur.

In each of the hardware chapters suggestions are given for things to make. We urge the student now, and again in the book itself, to design and build examples of the hardware. The essential aspect of technology is design; to understand design one must attempt it. In fact, another objective for the course we teach is to give the student the satisfaction of making something work.

A few notes on the presentation are needed. We use the euphemism "Third World" to refer to countries that are significantly less industrialized than the United States. Alternative phases — underdeveloped, developing, less developed, industrializing, poor, southern, and so forth — seemed either unwieldy or patronizing. A particular objection to "underdeveloped" is that many Third World nations have a culture at least as developed as that of the United States. We realize "Third World" is a very broad term used to include countries as different as Brazil and Nepal, but we hope that the context will make the meaning clear. When we use the unit of dollars we are referring to United States dollars exclusively so we do not use the abbreviation US $. Several of the examples use a local currency. We thought it misleading to translate uniformly into dollars because wage and price scales make such translations misleading for U.S. readers. We have generally used the units of length and weight that fit the example being discussed — feet and pounds in the United States and meters and kilograms elsewhere (see Table 1). We expect, in general, that most of our readers have United States backgrounds and we have usually written from that perspective.

Our intention was to make chapters as self-contained as possible, so not all chapters need be used. This results in some repetition. The electricity chapter, however, is background for the hydropower, wind power and photovoltaic chapters and the software chapters were written assuming familiarity with the hardware.

The book was written as a team effort and we deliberately retained differences in styles and formats, hoping this would improve the presentation. The team was composed of six undergraduates, a health care administrator, and two faculty, with a remarkably wide range of backgrounds. Members spent extensive periods of time in Bangladesh, Botswana, Chile, Jamaica, Kenya, Malawi, Peru, The Philippines, and Zambia. A world map follows this introduction.

A difficulty with preparing a book from lecture notes is that ideas read or heard several years ago subconsciously become incorporated and are not acknowledged. For this we apologize. We have used the book *More Other Homes and Garbage* by Jim Leckie *et al.* in our course and hope we have not lifted too much from it.

## Table 1. Units of Measurement

| | | |
|---|---|---|
| 1 acre | equals | $4.356 \times 10^4$ square feet |
| 1 Btu | " | 1054.8 joules |
| 1 cubic feet | " | 7.481 gallons |
| 1 cubic meter | " | 35.31 cubic feet |
| 1 foot | " | 30.48 centimeters |
| 1 gram | " | $3.527 \times 10^{-2}$ ounces |
| 1 hectare | " | 2.471 acres |
| 1 inch | " | 2.54 centimeters |
| 1 liter | " | 61.02 cubic inches |
| 1 liter | " | 0.2642 gallons |

Temperature in degrees Centigrade equals temperature in degrees Fahrenheit $-32$ and multiplied by 5/9

The abbreviation "kilo" means $1000 = 10^3$

The abbreviation "mega" means $1,000,000 = 10^6$

Many people made significant contributions to this book, besides those listed in our acknowledgments. We thank them all. Luisa Pacheco prepared an early version of the manuscript and organized that effort. Nearly all the noncredited pictures were taken by Mary F. Hazeltine, whom we thank sincerely. Technical assistance was given by Dr. Chris Case, the late Professor Joseph Loferski, Andrew Shapiro, Professor Robert Stickney, and Reginald Tewesa, to all of whom we are grateful. Shari Hirshman and Julie Iler helped us greatly during separate summers as undergraduate research assistants. It was a real pleasure to work with the professional staff at Academic Press: Joel Claypool, Julie Champagne, and Jane Phelan. The discussion in Chapter 13 benefited a great deal from interviews with people whose names are listed in Reference 8 of that chapter. We probably learned the most, though, from the students who took our course, and to them we are very appreciative.

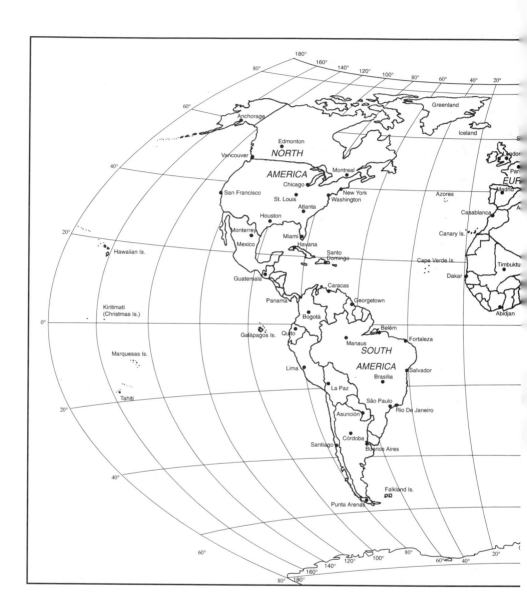

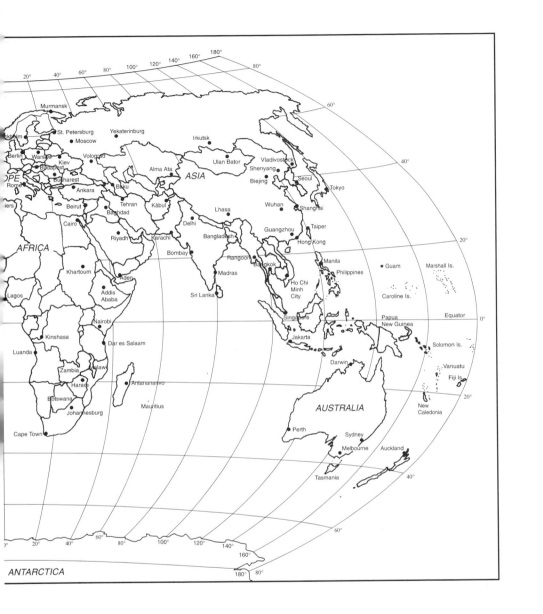

# APPROPRIATE TECHNOLOGY: WHAT IT IS AND WHY WE CARE

•  •  •  •  •  •  •  •  •  •  •  •  •  •  •

Made by a village Boy
from Eastern Zambia

**Reproduced with permission from the MIT Press from Arnold Pacey**
***The Culture of Technology*.**

## TECHNOLOGY CHOICE

All around us we see the results of technological choice. One way to renovate inner-city housing is to bulldoze it and rebuild with high-rise buildings. Another way is to train local people in carpentry and plumbing and make tools available to them. The first approach uses expensive equipment and complicated planning. The second approach involves simpler, cheaper tools and depends more on the judgment of the people doing the work. Both approaches require, for best results, imaginative and thorough thought. Which of these approaches — using sophisticated machines or using hammers and saws — makes the most sense?

Other examples of technological alternatives are readily apparent. Should the military buy a few very advanced bombers or a larger number of simpler ones if the cost is the same? The advanced bombers can do things simpler ones cannot but are harder to keep in working order. When I build an addition to my house should I integrate solar heating or should I install electric? A solar-heated room, with plants and a view, can be an attractive place, but without attention it will be too hot at midday and too cold at night. A thermostat on an electric baseboard heater is much more convenient. Should a farmer go the "hi-tech" route with mechanized tools, chemical fertilizers, and a single crop or should he/she try to grow several crops, using natural fertilizers whenever possible, and use only a standard tractor and implements? The first approach is probably more profitable when food prices are high but, because the equipment and supplies are usually paid from borrowed money, quite risky when prices fall. The second approach is probably less harmful to the environment in the long run but may not produce as much. In a sense, the owner-built greenhouse in Fig. 1.1 is a simpler technological alternative to agribusiness.

This book is about making technological choices of the kind described in the previous examples. To give the conclusion away, the answer to the questions from the examples is, "It depends." To make the choice sensibly one must look at what one is trying to accomplish, what expertise and resources are available, and what unintended consequences may ensue. (An unintended consequence of high-rise public housing may be concentrating people without cars far from jobs.) In other words, one has to decide which technology is most appropriate for a given situation. This book is therefore intended to provide some approaches to deciding what is appropriate.

**Fig. 1.1. Owner-built inner-city solar greenhouse.**

## APPROPRIATE TECHNOLOGY

What is "Appropriate Technology"? The preceding discussion implies that every type of technology might be appropriate somewhere, so in a trivial sense all technology could be appropriate. Such may be true, but the term Appropriate Technology is used in a narrower sense, and this book will focus on this narrower sense. The U.S. Congress's Office of Technology Assessment [4] characterizes Appropriate Technology as being small scale, energy efficient, environmentally sound, labor-intensive, and controlled by the local community. The Intermediate Technology Development Group [1], an organization which works toward the betterment of developing countries, uses nearly the same description but adds that the technology must be simple enough to be maintained by the people using it.

A central concept of Appropriate Technology is that the technology must match both the user and the need in complexity and scale. Some examples of how these two can be mismatched follow. A village in Botswana received 10 combines (grain-harvesting machines) as part of an aid project. When these machines broke down there were no trained mechanics and no spare parts. The combines lie rusting at the edge of the fields as the farmers use traditional methods to harvest their grain. The underpinnings necessary to support this equipment do not exist in the community. A small fishing community in the Marshall Islands receives a space-age photovoltaic system to provide electricity for lighting. Only after installation is it determined that there is no significant need for a system so sophisticated. The well-known tragedy caused by the explosion of a fertilizer factory in Bhopal, India, is another example of technology not matching local conditions, in this case presence of trained staff. In the United States, people point to the use of electric heating in houses, especially if such systems require new power plants, as a mismatch between a fairly simple need and an elaborate solution. One could say the same thing about one person using a full sized-automobile for commuting.

The proponents of Appropriate Technology, including the authors of this book, believe it is applicable to many situations both in the United States and in the so-called Third World. Certainly, it is an alternative approach that should be seriously considered. On the other hand, it does not make sense everywhere. Like other alternatives, it needs critical and serious evaluation.

## HISTORY OF APPROPRIATE TECHNOLOGY

Small-scale technology has been used for a long time, of course. The people who colonized America, for example, had no other options. The modern Appropriate Technology movement is attributed, however, to E. F. Schumacher, a British economist. Schumacher was an economist for the Coal Board and also an advisor to the government of Burma and later to that of India. He wrote several papers from 1955 to 1963, including "Economics in a Buddhist Country," "Non-Violent Economics," and "Levels of Technology," which were eventually incorporated in a book called *Small Is Beautiful* [5]. His ideas inspired the creation of the Intermediate Technology Development Group referred to earlier. The approach gained attention during the 1960s coincident with the social responsibility movements of those years. Several world leaders, including President Carter, became active supporters and many projects and support groups were formed. Some of these are still prospering in the 1990s, and their methods are certainly being used in many places. In the United States, the National Appropriate Technology Center offers information by telephone or mail, and in most states either government or private groups are doing relevant research or actually building projects. USAID, the foreign aid division of the State Department, now supports Appropriate Technology projects in many parts of the world. A colleague of Schumacher, George McRobie [3], wrote *Small Is Possible*, which reviews the history and status of the movement up to 1981.

Schumacher's focus was on jobs—people in the Third World desperately need employment. He argued that the capital investment required to create Western, high-technology jobs is too large to be practical. A modern foundry requires, perhaps, $20,000 for each job created. Existing technologies, such as those that a village blacksmith uses, correspond to an investment of $2 per job. An intermediate level is needed—for example, $200 per job, which is similar to a small machine shop in the United States. This intermediate technology could be much more productive than what now exists and could serve to move a country out of poverty without requiring huge capital investments. Schumacher further argued that the $200 technology would be close to existing methods and so would lead to improvement in skills for many local people. The $20,000 technology would be so technical that highly trained specialists would be required. The cost of training is high, so it would not be possible to train many specialists. Therefore, intermediate technology increases the knowledge level for most of the population, whereas high technology produces a class of experts separated from the rest of the economy. A final difficulty with $20,000 technology, such as an integrated circuit fabrication plant, is that it uses imported supplies, produces mostly export items, and needs expertise and components from outside the country for maintenance and improvement. Thus, the high-technology factory is not an effective way for the country as a whole to develop. The loom in Fig. 1.2 is an improvement over traditional looms. It was made by craftspeople at far less cost than imported automated looms and keeps many people employed.

**Fig. 1.2. Hand-made loom.**

# WHAT APPROPRIATE TECHNOLOGY OFFERS

A major reason for using Appropriate Technology, as Schumacher argues, is that it provides goods, services, and jobs that will not be provided any other way. No company or organization will be able to invest enough in high-technology factories in developing countries to provide sufficient jobs. An analogous argument is that the cost of tearing down old houses in U.S. cities and replacing them with modern high-rises is more than the government can afford. If a limited amount of money is available for low-cost housing, an Appropriate Technology approach, training people to rehabilitate their houses themselves, will have much greater impact. In the Third World few farmers can afford the equipment and chemical supplies to make farming similar to that in the United States but for a reasonable cost many farmers could be given the small machines and training necessary to improve their productivity.

A second advantage of Appropriate Technology is that it benefits most people, not just a few well-trained specialists. A concern for Third World planners is that Western technology only improves the lot of a small number of people — those able to work with it. Many of these people are already wealthy and educated. Western-style factories thus cause a rift between those working in the factories and those not and it is difficult for untrained people to move into the modern sector. If technology only slightly different from that existing is introduced, for instance, an improved farming method, then even those not directly involved with the new method can learn about it. More people would therefore benefit from the improvement. Many people could learn to make from scrap automobile parts the oxcart shown in Fig. 1.3, whereas fewer would gain expertise if carts were made in automated factories.

Introducing a new technology which is related to existing technology has two other advantages over a completely new technology. It is less disruptive to the social structure and it can be adapted. A factory or mine which brings young people from villages to a city affects village life much more than a small workshop located in the village. Augmenting the training of traditional doctors — so-called "herbalists" — often improves health care more than building Western-style hospitals because people feel more comfortable using herbalists and the cost is less. If a new technology is similar to the existing one, the user can adapt it most effectively to the local situation. High-efficiency wood stoves can be adapted by the user to burn straw or whatever fuel is available, whereas gas or electric stoves are much less flexible. Even methods of producing sophisticated equipment can be adapted. In one assembly plant in Taiwan, for example, pocket calculators

**Fig. 1.3. Improved oxcart. (Drawing by Kwesi Davis.)**

are produced not on assembly lines but by small groups working in teams, just as people worked together in traditional society. The process is more cost effective than U.S. production methods.

Appropriate Technology benefits a society more than high-technology because it fosters self-reliance and responsibility. Compare a person who has renovated his/her own house with one who has been assigned an apartment. Who is probably more prepared to take responsibility in a job? Similarly, an experienced owner of a small farm is probably better prepared for a leadership role in the community than a worker on a mechanized farm. Appropriate Technology not only teaches skills, but also gives people experience in solving problems and getting things done. Many people, such as the tinsmith in Fig. 1.4, would much prefer to be their own boss, to be responsible for themselves, than to work for another. Appropriate Technology is more likely than high technology to provide this opportunity because smaller organizations mean more organizations, with more leaders.

This same quality, of being responsible for one's own success, also applies to groups. When high technology is brought into a country or a region it is often implemented by a large company. Decisions about the factory, such as how it is run, or even whether it will continue to exist if the market changes, are made at corporate headquarters. The result is that the people working in the factory do not have control over what happens. They may recognize ways to improve the process or the end product but will have difficulty getting corporate management to pay attention. Appropriate Technology is small scale and thus the people directly involved can have

**Fig. 1.4. Tinsmith in Singapore.**

significant control. A small machine shop with general-purpose machines can adjust its products and meet market needs more effectively than an automated factory, partly because the tools are more adaptable and partly because fewer people have to be convinced in a small group. Community self-help groups are often more effective at meeting housing needs than bigger, citywide organizations because the community groups are closer to the people and their needs. Of course, situations do exist in which large size is essential; it is hard to think of an economic car factory which is small, and a small group of artisans making pottery may need a large organization to do their marketing because it may be difficult for a small company to gain access to a market.

Advocates of Appropriate Technology indicate that not only does it foster self-reliance and responsibility but it also fosters other desirable attitudes, including cooperation and frugality. If machines are not available, then people's strength must substitute, and this usually means that groups of people must work together. In colonial United States the neighboring farmers would gather to help put the roof on a new house. Many of the successful Appropriate Technology urban projects have been done by neighborhood associations or cooperatives. The reason Appropriate Technology promotes frugality may be simply that the users have less to waste or it may be that hand work encourages care and concern, which translates into less waste.

A related reason why some people are attracted to Appropriate Technology is the *type* of job it produces. It is not surprising that a person operating a self-sufficient farm, (Fig. 1.5) is willing to work 55 hours a week, whereas an assembly-line worker feels 40 hours are too

**Fig. 1.5. Subsistence farm in the United States.**

**Fig. 1.6. Pottery factory in Malawi.**

many. The farmer has more interesting, challenging, and rewarding work. The satisfaction one gains from seeing a farm succeed is, for some people, worth much more than the money one could earn in another job. Similar feelings of accomplishment come from other Appropriate Technology activities, as demonstrated by the pottery shown in Fig. 1.6. The types of jobs promoted by Appropriate Technology are easy to integrate into a lifestyle that many people aspire to — self reliance, fulfillment through one's profession, and concern for others.

A final set of advantages for the Appropriate Technology approach is related to the environment. Appropriate Technology farming, for example, minimizes the use of chemical fertilizers and pesticides. It uses manure and other waste products as fertilizer. It avoids the use of heavy machinery which damages the ground. It entails growing several different crops at once. All these help to preserve the soil and the surrounding waterways. Solar energy reduces the amount of fuel burnt to heat a home, so the air is less polluted. Craftpersons working in small groups are often more careful than unskilled workers on an assembly line, so small-scale hand production may generate less waste than mass production. Appropriate Technology can often be less of a burden to the environment than other technologies. One must, however, be careful with this argument. Use of a wood stove would seem to offer many of the advantages of Appropriate Technology, but if most people used wood stoves, air pollution would be serious and the forests would be severely depleted. Few people recall that a major reason automobiles were originally encouraged was to reduce the serious pollution problem caused by horses in city streets. Now most people would consider automobiles much more of an environmental threat than horses.

A related environmental concern is diversity. So-called "modern" agriculture consists of planting a single crop using the most productive seed variety. A plant disease endemic to that variety then can wipe out a whole season's yield for an entire region. Such came close to happening with corn in the United States in 1974. The Appropriate Technology approach resorts to a variety of crops, so it is less likely a disease would spread widely, and a diversity of species, so at least some should survive a blight.

## CONCERNS ABOUT THE USE OF APPROPRIATE TECHNOLOGY

It is worthwhile to consider why Appropriate Technology may not be a desirable approach either in the United States or in the Third World. As various technologies are described in later chapters, the reader can evaluate these objections. In any case, the authors are not claiming that Appropriate Technology is useful in every situation.

A major concern is whether Appropriate Technology can provide sufficient goods and services. It does not appear that small-scale hydroelectric systems would by themselves give sufficient energy for the United States. Could small farms provide enough food? Could small factories produce at sufficiently low cost or are they inherently always inefficient compared to mass production? In later chapters we will consider these questions and try to determine when Appropriate Technology really does make sense.

A related concern is whether people will accept the Appropriate Technology approach or the inventions designed by Appropriate Technology. Even if oil heating costs were higher than solar, would people prefer to use oil? Do many people really prefer a high-technology solution such as a complex kitchen gadget to a more effective lower tech one? (The Defense Department is accused of this bias toward high tech.) Would most people really prefer to work 40 hours a week at a boring job and then pay someone else to improve their house even though they have leisure time or would they prefer to do their own work if they could?

A concern for Third World countries has been whether Appropriate Technology does in fact lead to a national development, in the sense of a trained workforce. Will a nation following the Appropriate Technology approach end up with a population generally conversant with technology, but with a technology that is of no use in the modern world? Will people with a background in Appropriate Technology have an easier time learning newer technologies compared to people with no background at all? An analogous question in the United States is whether people trained to rebuild their own homes can use those skills elsewhere.

Some Third World leaders are understandably suspicious of people from the industrial countries and wonder if Appropriate Technology is a way to discourage the Third World from industrializing and becoming competitors. One answer to these leaders is that there may be no other way for the nations of the Third World to industrialize on their own.

Another area of concern is that the Appropriate Technology approach seems to be difficult to plan or manage. Would it not be better to use established technologies from the United States, for example, and spend one's time and effort making them work? By definition, Appropriate Technology is specific to a locality, so transferring expertise from one country to another is difficult. Because Appropriate Technology is small scale and done by many independent people, it is difficult for a government official to understand what is happening and take action when needed. People who object to Appropriate Technology for this reason, however, may be too optimistic about the problems in transferring an established technology and too pessimistic about the problems in transferring Appropriate Technology.

## CRITERIA FOR DECIDING IF A TECHNOLOGY IS APPROPRIATE

In light of the above discussion, what general guidelines can be given for deciding if a certain technological approach should be used? Three kinds of questions seem valid: (i) Does the technology provide the goods and services it must at a reasonable cost, including long-term costs?; (ii) Does it have a desirable influence on the local culture now and will it in the future?; and (iii) Does it promote a healthy lifestyle for the individual?

Part of the cost–benefit analysis is easy and part is difficult. Calculating or estimating the direct outputs of a technological choice is usually straightforward and we will do much of it later. Direct cost calculations are also usually straightforward. We can estimate how many fish we will get from a pond and what it costs to make and maintain the pond. Long-term effects may be harder to foresee and to account for. What are the real costs, in 30 years, of petroleum depletion? Petroleum is not only a fuel but also a raw material for many plastics. How do I put a cost on an endangered plant whose habitat is destroyed by the pond I built for hydroelectric power, especially if I think the plant looks unattractive? No definite rules exist to answer these questions. The designer must use her/his own judgment.

Again, the designer must make judgments on the magnitude of the effect a new technology will have on the culture of local people. The designer must also judge if that effect is good or bad. One thing does seem to be true: Rapid technological changes almost always weaken an existing culture. People need a sense of belonging to a group with shared heritage and values and it takes time to rebuild that shared heritage once it is broken. A problem for anyone working in another culture is to determine how much one tries to make changes in that culture to meet one's own standards. If one feels that a new technology introduced will worsen the lot of the elderly, minorities, or women, what does one do — especially if the people involved do not seem concerned? How does one react if local people make consequences that seem shortsighted (e.g., chemical factory near a wildlife preserve and a record player rather than a water pump)?

Another issue in evaluating a technological choice is consideration of the future. If a particular choice is made now, what capability and needs — both expertise and infrastructure (roads, electric power, government, and so forth) — will be available in 10 years? What kinds of lifestyles will be possible? While the designer may have difficulty foreseeing the future decisions, he or she must at least lay out the options and their possible consequences.

## PLAN FOR THIS BOOK

The book is divided into two parts. The first, Chapters 2–12, focuses on the technologies themselves. The second part, Chapters 13–17, focuses on how these technologies are used. The theme of the first part is design: How does one design a device or a system to meet a need? The theme of the second part is policy: What policies — governmental or private — will result in a sensible choice of devices or systems and their effective implementation? As we will see, in many cases the policy questions are the most difficult. Many examples exist of new technologies that work well but do not make life better for the people involved. To describe the plan another way, the question is really in making choices. The first part describes the technological choices. The second part describes what choices must be made to make a particular technology acceptable and fit the needs of the people applying it.

The reason we start with the technologies is that one cannot really make a sensible decision about using a tool or a new material unless one understands how it works. It is dangerous to depend on experts to advise us if we do not have the background to evaluate their advice. This book starts with a review of the design process (Chapter 2) and then provides an introduction to electricity (Chapter 3) as used in the home to power appliances and electronics. Electricity can be generated by waterwheels (Chapter 4), by wind machines (Chapter 5), or by solar cells (Chapter 7). Small but effective machinery, such as efficient wood stoves, bicycle-powered devices, and water pumps, are the topic of Chapter 6. A way of converting manure and other vegetable wastes to gas useful for cooking is considered in Chapter 8. Food production without expensive machinery is discussed in the next two chapters: Chapter 9 discusses growing vegetables and raising small animals and Chapter 10 discusses fish farming. Next, we will look at alternative approaches to health care. Finally, in Chapter 12, we examine how solar energy can be used to heat a building. These particular technologies were chosen because, on the one hand, they are important both in the Third World and in the United States, and, on the other hand, they can be understood and used by people without a strong engineering or science background.

The policy chapters describe what has happened when new technologies have been introduced and give guidelines on how an introduction can be effective. We begin with the Third World (Chapter 13) because the issues seem clearer, perhaps because we know less about the real situation. In Chapter 14, we consider the United States. Does Appropriate Technology have something to contribute here? Chapter 15 deals with the effect of technological choices on the culture of the recipients of the new technology. An important cultural aspect has been the different effects on women and men. Chapter 16 discusses policy making and Chapter 17 considers the future of Appropriate Technology.

Each chapter contains a set of problems and projects. One reason for doing problems, of course, is that they help in the learning process. Another reason is that technology consists of actually doing something — of action, not just contemplation. In order to understand what a designer faces, one must try to design. A person does not learn to drive a car or play tennis or write poetry by reading about it. Some of the projects involve actual construction — building a solar hot water heater or growing lettuce without soil. Other projects are systems design — planning a fish pond or a public health campaign. Both kinds of projects are meant to require analysis of costs and benefits, of unintended consequences, and of exactly what must be done step by step. Only by going through the process of thinking out the whole project does one really understand how technological change occurs. An example of a student project in our course is shown in Fig. 1.7.

**Fig. 1.7.a. Preparing to cut an oil drum.**

**Fig. 1.7.b. Completed vertical axis wind machine.**

# USEFULNESS

The reader may be asking how the material in this book is of any real use. Why should one care? An answer, as is probably obvious, is that the authors believe Appropriate Technology has much to offer both the Third World and the United States.

We believe the material is useful in a variety of other ways. First, the subject matter—food, energy, health care, Third World countries, cultural change and others—is interesting in itself, and at least some of the subjects will be important in the daily life of the reader. Keeping a house warm without going into debt is not always a trivial matter, and many readers will be faced with analyzing the impact of a major technological decision in their community, such as solid waste disposal or houses filling up farmland. The issues are also important. How should technological choices be made? How will those choices change the lives of the people involved? Many people, in their jobs, will be concerned about whether the so-called developing countries will survive or whether they will be so successful as to become serious competitors. Knowledge of technological change and technology transfer, which we will consider, is helpful in this regard. While Appropriate Technology itself may not be immediate to every reader, study of it is a useful way to learn about things that are immediate.

Finally, technology can be fun. It is exciting and satisfying to design something and see it work. Appropriate Technology is a good place to start because the technologies can be mastered relatively easily. From their course, the authors have been gratified to see how much students can accomplish both in design and in construction, and how much excitement and satisfaction the accomplishment brings. We are confident the readers of this book can do similar things. Satisfaction comes in many ways besides creative work itself. One other way is through being of help to others, and Appropriate Technology may be a way of being helpful. Our last point is that while technology may not be the entire solution to the problems facing the world, it is an essential part of most solutions. In other words, if one hopes to contribute to the betterment of life for others, it is most helpful to understand aspects of technology and how these aspects influence society.

# REFERENCES

1. *Appropriate Technology*, A journal published by the Intermediate Technology Development Group, 9 King Street, London WCZE 8HN.

   *Includes both technical articles and philosophical ones — primarily a Third World focus. The editorials, especially, give insight into what the Intermediate Technology Development Group is about.*

2. Florman, S. (1976). *The Existential Pleasures of Engineering*. St. Martins, New York.

   *Why technology is fun. Florman is not a supporter of Appropriate Technology so his book complements this book as well as the other references listed below.*

3. McRobie, G. (1981). *Small Is Possible*. Harper & Row, New York.

   *An update to 1981 on the appropriate technology movement. A list of places, with addresses, where projects are being conducted. Good history of what has happened and is happening.*

4. Office of Technology Assessment (1981). *An Assessment of Technology for Local Development*, (GPO Stock No. 052-003-00797-5). U.S. Government Printing Office, Washington, DC.

   *Focuses on the United States and describes successful projects such as solar installations. Readable and inspiring but does not include much on design.*

5. Schumacher, E. F. (1973). *Small Is Beautiful*. Harper & Row, New York.

   *Much of this is difficult and requires some background in economics. In our course, we suggest starting with Chapter 3 of Part III, "Social and Economic Problems Calling for the Development of Intermediate Technology." Schumacher, of course, is the father of the movement.*

# PROBLEMS AND PROJECTS

1. Consider a major national problem, e.g., defense, solid waste disposal, and the homeless. Devise both a high-technology approach to its solution and an Appropriate Technology approach.

2. Some people have written that the use of computers in teaching is really a mismatch — a technical solution to a human problem. Do you agree? Suggest alternative ways of teaching which are consistent with Appropriate Technology ideas.

3. Consider a specific group of unemployed or underemployed people. How could a small-scale technology approach create jobs for them?

4. Why does one not see or hear more about self-help housing renovation projects? What might be done to foster such projects?

5. Small autonomous groups are part of the way Appropriate Technology is usually done. Could this emphasis on small autonomous groups improve the way organizations doing other things function? Give an example of how other kinds of work might be reorganized using this approach.

6. Appropriate Technology alternatives do not seem to exist in most of the communications (e.g., telephone, radio, and television) and computer industries. Why not? What seem to be the characteristics of industries in which Appropriate Technology does give a useful alternative?

7. Consider examples of small-scale technology from U.S. colonial history. Are such technologies still in use? If not, why not?

8. Consider an environmental problem (e.g., acid rain and polluted rivers) important to your community. Could an Appropriate Technology solution be useful? How?

9. What would it take to induce people to commute by bicycle rather than by car? (For example, jobs near homes, safe bike paths, expensive gasoline, and much advertising.)

10. How could a state government in the United States encourage the use of an Appropriate Technology (e.g., wind generators on farms)?

11. How would you decide if it were worthwhile to build a pottery factory in a remote African village to produce hand-made dishes? Would the dishes be sold either in urban areas of the same country or for export? (See Fig. 1.6.)

12. Consider three important national issues and identify the technological component. For example, insider trading on Wall Street is exacerbated by rapid and accurate electronic communications.

# 2

# BASICS

Chart of Standard Business Procedures

| 1 As sales requested it | 2 As production ordered it | 3 As Engineering designed it | 4 As the lawyers approved it |
| 5 As Management approved it | 6 As Advertising promised it | 7 As the Plant installed it | 8 What the customer wanted |

Three different subjects are discussed in this chapter. Each is basic to understanding and using technology. First, we describe how new technology is created. Second, a technique for comparing the costs of alternatives is discussed. Finally, energy fundamentals are considered.

## WHAT DESIGN IS

This section is intended to give some insight into the design process. Actually, the entire book is about design. We will consider in later chapters how to design electrical systems driven by wind or water power, how to design houses that use only a little heating fuel, and how to design farms growing vegetables or fish. Engineers design, which is what distinguishes them from scientists. Of course, other people also design — architects, tailors, health care planners, and many others. The designer, in every case, starts with requirements and ends up with something that meets those requirements. The designer gets to see her or his ideas actually built, and actually meet a goal.

## *An Example of the Design Process*

Three designers were asked to submit proposals for a short-haul (10-mile) transportation system to be built by a community of 100,000. The ridership would be approximately 30,000 trips per day with an average distance of 4 miles. The coverage is to be radial, with lightest concentration at the perimeter and heaviest at city center.

One designer developed an automobile-based system. Another came up with a bicycle system and the third presented a subway system. Each proposal was thorough, professional, and presented a strong case.

Before we present the decision of the community on which design is to be implemented, let us look at the process the designers went through to arrive at their solutions. More about the process can be found in many engineering books [2]. The steps in the process are shown in Fig. 2.1, which shows the steps each designer went through for our community.

When the proposals were first requested the designers had many questions. Who are the riders? What is the terrain? Is there anything existing that might be incorporated? What is the climate? What should the cost be? When is the system needed? What sort of lifetime is expected? What construction techniques are available?

The first task, asking questions, is to clarify the problem so the designers can know what solutions are reasonable. It usually happens that, unless the same person is the formulator of the problem and the provider of the solution, initially the problem is not well enough understood to give a solid starting point.

Once the problem is well-defined the designer has the information to develop an initial concept for the design. The step from problem to concept is the most difficult to quantify. To this step designers bring their own experience, culture, training, and personality. The idea behind the initial concept may come from having seen a similar situation, from a sudden flash of inspiration, or from a gradual synthesis of parts into a working whole. In our example, three different designers came up with three different concepts: automobiles, bicycles, and subways.

After the hurdle to concept has been made, the testing begins. Following are some general criteria that most designs must meet to be viable:

*Effectiveness*: The concept must meet the requirements. A self-sufficient farm must grow enough food to feed its inhabitants.

*Reliability*: A reliable product performs as advertised when called upon. Measures of reliability are frequency of repair and mean time before failure. The bicycle system fails on snowy days. Some cars do not start on cold mornings.

*Maintainability*: Nearly every device needs maintenance or upkeep. The fewer maintenance problems the better. The user or service person should be able to perform routine maintenance easily and safely. In an auto this might mean designing the engine compartment for easy access to the spark plugs. In a village water pump it would mean making the seals accessible with common hand tools.

*Reparability*: Should the product fail, the failed part should be replaceable and the task of replacing it should take only a reasonable amount of time and effort. Imported machines in Third World countries stand idle because the parts needed for repairs are not there.

*Availability*: The product should be functional when you need it; that is, your car should start when you turn the key. Subways may not be available during a power outage. Designs must deal with the availability of their energy source.

Failure to meet any of these criteria is usually fatal to the concept. Back to square one, as shown in Fig. 2.1. We might have to go from initial concept to test and back again several times, trying different initial concepts each time.

If the concept passes the test—that is, it meets each criteria—then a preliminary design is made and again evaluated. In our automobile system, for example, the location of roads and the construction of parking facilities would be considered in making the preliminary design. The evaluation here asks two kinds of questions of the preliminary design. (Only one block is shown in Fig. 2.1 for both sets of questions.) The first set of questions might be called hardware related:

- What is the worst case combination of high demand and low availability and how would the design perform in this worst case? What grade of service would a subway give on the Friday before a long weekend if two subway cars were in the shop for repair?

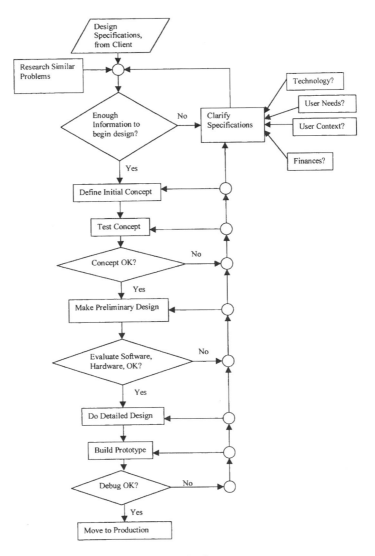

**Fig. 2.1. The design process.**

- Is the necessary material available?
- Is the labor for construction available?
- What is the monetary cost?
- Is the construction time reasonable?

The second approach to evaluation might be called software related because it deals more directly with so-called "soft costs" — those hard to put a number on:

- What is the environmental impact?
- What is the effect on the user's lifestyle?
- Does the design exclude any potential users?
- Does it provide a path for adapting to changing needs?

Failure at the preliminary design level may not be fatal because this design can be substantially modified. Of course, a single design is almost never best in all regards; a short construction time, for example, may mean a large cost. The designer must make trade-offs.

Once the questions in the two sets and similar ones have been answered, the rest of the design is in doing the details. However, these details are important; every nut and bolt must be specified. As much as possible, every design decision must be made ahead of time. The reason for this is twofold. The designers usually do not execute the design. Therefore, they must communicate it to the craftspeople who will execute the design based on drawings, numbers, and words. The detailing process may also bring to light subtle flaws, such as a misplaced hole or a short circuit, that are much easier to correct on paper than in concrete or silicon. In some cases, however, some design decisions must be made in the field, that is, during actual building. In highway construction, for example, the ground may turn out to be entirely different from expected. The detailed design itself must also be studied carefully and modified if necessary before actual construction.

In the final stage a prototype or model is constructed. Ideally, this is a model that may be tested under working conditions. It is almost impossible to foresee and design for every contingency. The prototype gives the designer one more chance to test the concept before committing it to production.

A working prototype signals the success of the designer's effort. The final step is to see the design into production.

## Major Points about Design

The preceding discussion of necessity, glossed over many issues that play a significant part in the design process. The following major points should be made:

- Design is an iterative process. One may go around the loops shown in Fig. 2.1 many times until a concept, preliminary design, or detailed design is satisfactory. The problem does not go away if one solution fails. The rejection of one solution means looking for another. If the design fails some test it is modified until it passes the test or fails fatally.

- As a design proceeds it becomes more difficult to reject its basic concept and start again. The investments of time, energy, psyche, and capital give a well-along design too great a value to be dropped. The implication is that it is better to spend much time working on the original concept than to try to salvage a poor idea in the final phases.

- An "ideal" solution rarely sees the light of day. There are too many competing constraints. Trade-offs must be made to get some semblance of the initial concept into production.

The community in our example selected the subway system. The subway was not chosen because it was ideal. The other proposals were rejected because they were more flawed. Bicycles did not offer enough protection in poor weather and a significant portion of the population could not use them. Autos consumed too much land and posed a greater threat to the environment.

These reasons for rejection of a design had been considered by the respective designers. Each felt that her/his design had enough merit to outweigh its flaws. The rejection of two designs illustrates that no matter how "good" a designer feels the product to be, the ultimate test is in the market. If the consumer is not willing to give something of value in exchange for the product, the design is a failure.

The second point discussed previously — the difficulty of changing an established idea — is worth considering further because it applies not only to design. Once a technology is in use it is very hard to scrap it and start over. Color television was developed after black-and-white TV. The color system had to be designed so black-and-white sets could still be used although a better color picture could have been obtained by starting from scratch. Barry Commoner [3], a noted environmental scientist, has pointed out that pollution from automobiles would be much less if a whole new system, such as electric-powered cars, had been chosen because emission controls currently in use are complicated and expensive. As in many situations in real life, it is much better to prevent a problem during the design stage than to fix it later.

## *Problems with Evaluating a Design*

The first problem with evaluating a design is balancing tangible and intangible factors — the hardware and software questions described previously. A tangible factor is how much a bicycle path costs. An intangible factor is how attractive the path can be made by preserving trees and wildflowers. In the United States people are accustomed to comparing alternatives using monetary terms, so things that are difficult to put a price on are often considered less important. This problem of how to deal with factors that are hard to quantify is considered in Chapter 16 when we consider policy making. An approach to solving it is simply to be as careful as possible to take intangible factors into account when designing.

A second problem is more worrisome. In many cases the people who benefit from a design are different from those who lose from it, so it is difficult to make fair trade-offs. The people who benefit from a new highway are often those who live in suburbs and can commute easily. Some of the people who suffer are those whose homes are adjacent to the new highway, and some of these may not live far enough out of the city to need the new highway.

Chlorofluorocarbons used in cans of spray paint as a vehicle are intentionally made inert and light in weight in order to produce a smooth painted surface. The result, however, is that after the paint is deposited they do not combine with other substances but float up into the stratosphere. There solar radiation, stronger than on the earth's surface, decomposes the chlorofluorocarbons and the decomposed chemicals deplete the ozone in the stratosphere. However, it is the ozone which protects people on the earth's surface from harmful solar radiation. The concern, of course, is that future generations will be worse off, that is, be more susceptible to cancer, because of the convenience given to this generation by spray cans.

It has been very difficult to find fair ways of taking into account the interests of people affected by a decision made by someone else. Doing justice implies that someone or something, perhaps the government, represent these interests; however, this representation has not always been effective in the past. Perhaps the best that can be done is simply to urge the designer to think carefully about people who might be affected by a design decision. The weight of doing justice therefore falls on the designer. There are no simple solutions.

An important issue in any design is cost. In the next section the method used in calculating costs will be described.

# CAPITAL RECOVERY FACTOR

Assume we wanted to compare the costs of two projects; one with a high initial cost but low annual costs and the other with a low initial cost but a high annual cost. An example would be heating a workroom. One alternative would be passive solar: big windows and heat storage. This would cost $1000 to build but annual costs would be only $50 for cleaning and caulking. The other alternative is a kerosene heater. The initial cost is lower ($100), but the annual cost, mostly for kerosene, is much higher ($250). Which of these is cheaper if the system must last for 10 years? The problem is made difficult because one probably would borrow the $1000 to build the solar system and so must pay back interest as well as the original amount. Without interest charges the passive solar system would cost $1500 (1000 + 50 × 10) and the other $2600 (100 + 250 × 10).

A convenient and valid way to make the comparison uses *the capital recovery factor*. The capital recovery factor gives the amount of money that must be repaid each year if the initial amount is borrowed. Of course, the factor depends on the duration of the loan and the interest rate. The values of the capital recovery factor are shown in Table 2.1. A complicated formula can be used to find these factors but the table is easier. Most spreadsheet programs give these factors. The table shows, in the upper left-hand corner, that if we borrowed $100 for 3 years at 5% interest we would have to pay back $36.70 each year. On the other hand, the lower right-hand corner shows that a $200 loan at 20% for 20 years requires annual payments of $41.00. To get this annual payment, we multiply the value of the loan by the capital recovery factor:

$$200 \cdot 0.205 = \$41.00 \qquad [2.1]$$

### Table 2.1.  Capital Recovery Factors

| Years | Interest rate (%) | | | | | | |
|---|---|---|---|---|---|---|---|
| | 5 | 8 | 10 | 12 | 14 | 15 | 20 |
| 3 | 0.367 | 0.388 | 0.402 | 0.416 | 0.431 | 0.438 | 0.475 |
| 5 | 0.231 | 0.250 | 0.264 | 0.277 | 0.291 | 0.298 | 0.334 |
| 7 | 0.173 | 0.192 | 0.205 | 0.219 | 0.233 | 0.240 | 0.277 |
| 9 | 0.141 | 0.160 | 0.174 | 0.188 | 0.202 | 0.210 | 0.250 |
| 10 | 0.130 | 0.149 | 0.163 | 0.177 | 0.192 | 0.199 | 0.239 |
| 15 | 0.096 | 0.117 | 0.131 | 0.147 | 0.163 | 0.171 | 0.214 |
| 20 | 0.080 | 0.102 | 0.117 | 0.134 | 0.151 | 0.160 | 0.205 |

Banks use similar tables to compute mortgage payments for homes or car payments, but these have monthly payments. In the examples in this book such precision is not necessary.

How does one know which interest rate or which number of years to use in an actual situation? If one really would borrow the money from a bank, then the rate to use, of course, is what the bank charges. If one takes the money out of a savings account, then the interest one would have earned should be used in the calculations. Even if one does not know where the money would come from it is reasonable to use the interest rate charged by a local bank. Regarding deciding the number of years to use, one should use the actual number of years if a real loan is taken. If a real loan is not taken then the life of the equipment is a good choice because one often wants to be paying for equipment for about as long as it is useful. (Of course, estimating the life of equipment may not be easy either.)

Now let's compare the two ways of heating the workroom mentioned earlier. Let us assume the relevant interest rate is 12%. Let us also assume both our solar system and the kerosene heaters last for 10 years. The capital recovery factor is 0.177 (see the fifth row and fourth column of Table 2.1). The annual cost of the solar system has two parts. The first, sometimes called the capital amortization cost, equals the initial cost ($1000), times the capital recovery factor. The second is the annual maintenance costs ($50). The total annual cost is $227.00. The money spent paying back the large initial expense is the greater part of this annual cost.

- *Annual payments: 1000 × 0.177 = $177.00*
- *Maintenance*                  *50.00*
- *Total annual costs*         *$227.00*

The annual costs for the kerosene heater are calculated as follows:

- *Annual payments: 100 × 0.177 = $17.70*
- *Fuel and maintenance*        *250.00*
- *Total annual costs*         *$267.70*

In this case, the solar system is cheaper, although the annual costs are similar.

To summarize, one compares projects with different initial costs and different annual costs by computing a total annual cost. Part of the annual cost is derived from the initial cost by using the capital recovery factor. This part of the annual cost is the principal and interest payments. The rest of the annual cost is out-of-pocket expenses, usually supplies and maintenance.

This financial analysis can be made more accurate in several ways, such as accounting for salvage value. In our example, the salvage values are the worth of the solar system after 10 years and the worth of the 10-year-old kerosene heater. Another refinement is accounting for taxes, which are different for capital improvements such as solar greenhouses than for operating costs such as fuel. A third refinement is trying to account for inflation. We will not use such refinements in this book. Standard engineering economy books such as the one by Riggs and West [4] can be helpful and provide in-depth analysis.

# ENERGY CONVERSION

This is a discussion of energy conversion. Its purpose is to explain some of the ideas that are common to the following chapters on technology. Although the chapters on hydropower, wind power, tools, photovoltaic, methane digesters, agriculture, aquaculture, and solar energy present very different technologies, there are some underlying concepts that are shared. A good understanding of these common ideas will make it easier to grasp their different implementations.

## *Science and Technology*

Science is our understanding of the nature of things. People practicing science develop and test theories about the way the universe functions. The application of our scientific knowledge to practical needs is technology. Technology and science are not necessarily disjointed; each may lead the other through the development process. It is interesting to note that although theory may change, the technologies applying it are little affected.

These days leading-edge technology comes in two forms: information, which collects and manipulates data, educates, entertains, and informs, and energy technology, which accomplishes physical tasks such as making a product or moving goods and materials. These forms may have components of each other but their purposes distinguish them. We will concentrate on energy.

## *What Is Energy?*

Energy is the capacity for doing work. The energy in a flowing stream corresponds to the work we could accomplish if we used the stream to drive a waterwheel. The energy in a battery represents the work an electric motor would do if connected to the battery.

## Energy Comes in Many Forms

Light from the sun, water flowing through a river, wind blowing, and electricity flowing through wires are all forms of energy. Using a water-wheel we can harness the energy of the flowing river to help us grind grain or produce electricity. A windmill can use the energy in the wind to pump water for irrigation. A purpose of technology is to change one form of energy to another. In changing, the energy performs some task.

Energy comes in many forms, each with its own constraints. The most basic distinction is between potential and kinetic energy. Potential energy is not yet active. A coiled spring, a ball held above a table, a battery when not in use, and a lump of coal all possess potential energy. Kinetic energy comes from the motion of things. Releasing the spring, dropping the ball, putting the battery in a circuit, and burning the coal transform the stored potential into kinetic energy. In the case of the battery the things in motion are small bits of electricity called electrons, which flow through the circuit. In the case of the burning coal the things in motion are the molecules heated by the burning coal.

Another way to divide energies is by considering the way they travel. Mechanical energy is transmitted through gears, shafts, belts and pulleys, chains and cogs, and connecting rods and pins; generally anything that can move. Electrical energy travels most easily along wires. Light, which is actually akin to electric energy, can be transmitted through any transparent or translucent material, gas, or vacuum. Heat will travel through anything — solids, gases, and vacuum. Fluid energy, obviously, travels with the moving fluid. Chemical energy, ready to be released by chemical reactions, travels in containers or in materials.

It is easy to blur these distinctions by turning our theoretical all-powerful microscope on them. A fluid viewed at the molecular level can be considered a mechanical system with particles bumping against one another. The same might be said of electrons traveling through a wire. The point is, if we break energy into logical divisions, understanding and manipulation come more easily. Table 2.2 shows various forms of energy.

## Energy Can Be Converted from One Form to Another

The purpose of technology is to aid us in accomplishing a task. The task may be drilling a hole in a piece of wood, typing a paper, or traveling across the country. To drill a hole we might use a brace and bit, a sharp stone and a bowed stick, an electric drill, or a laser. No matter what tool we use the result basically is the same.

## Table 2.2. Forms of Energy

| Form | Potential | Kinetic |
|------|-----------|---------|
| Mechanical | Wound spring | Driven watch |
| Light | None | Light |
| Chemical | Battery | None |
| Electrical | Charge in a capacitor | Current through a wire |
| Fluid | Water behind a dam | Wind |

Intrinsic to each of these tools is energy conversion. Take, for instance, the electric drill. To make a hole in wood the drill takes electrical energy from an electric outlet and, by passing it through some coils of wire near a magnet (do not worry about the details now), converts the electrical energy to the rotating motion (or mechanical energy) of the drill. The sharp edge of the drill bit cuts the wood. Both the cutting and the rubbing of the bit against the wood convert the mechanical energy to heat.

It is not interesting that we have various forms of energy, but it is interesting that we have developed ways to convert one form to another. Through the conversion process we can perform useful work. The history of technology chronicles the inventions of different methods of energy conversion. Fire converts chemical energy to heat and light. A wheel converts linear mechanical motion to rotary motion. Waterwheels convert fluid energy to mechanical energy. Steam engines convert chemical energy to mechanical energy. Electric lights convert electrical energy to light and heat. Table 2.3 lists some processes that convert from one form of energy to another.

With the growth of energy-conversion techniques came the growth of theory to explain and quantify the methods. Delving into the theory will give us a common ground for discussion of various technologies.

## Table 2.3. Conversion Processes

| From \ To | Mechanical | Electrical | Light | Chemical |
|-----------|------------|------------|-------|----------|
| Mechanical | Wheel | Generator | None | Heat Pump |
| Electrical | Motor | Wire | Light bulb | Charge battery |
| Light | None | Photovoltaic | Fiber optics | Biomass |
| Chemical | Heat engine | Discharge battery | Combustion | Methane Digester |

## Some Theory

To accomplish tasks we do work. Strictly defined, work is the transfer of energy from one physical system to another. An engineer's basic example of work is pushing a block across a level surface. In doing this we transfer energy from our body to several kinds of energy. When the block is in motion it possesses kinetic energy. The friction of the block against the surface converts mechanical energy to heat. Work is quantified by multiplying the force applied by the distance traveled—a force of 1 pound pushing for a distance of 1 foot uses 1 foot-pound of energy.

The following are some useful facts about energy: Energy is neither created nor destroyed. This is a cosmic fact, true everywhere in the universe. Our intuition tells us that we consume energy when we drill a hole (do work). In reality, the energy is converted to other forms, chiefly heat, and the total amount of energy in the universe remains fixed. If this is true then why can't we keep using the same energy over and over? Because the form of the energy is changed in the conversion—usually to one less useful.

It is this conversion process upon which we will be stubbing our toes. Our interest is in the amount of useful work that can be done. The useful work must be distinguished from the total work. In all conversions some of the energy is converted to heat. The reason it seems as if we consume energy is that the heat is transferred to the surroundings and performs no useful task (unless we are trying to heat something).

## Energy Can Be Transferred

Energy can be transferred. Wires transfer electric energy from one place to another. Electromagnetic waves transfer light, heat, and other forms from one place to another. Heating ducts transfer heat in the form of warm air from a heated room to a cool one. The axle on a waterwheel transfers energy from the wheel to the electric generator or to the stones that grind the flour. The problem in transferring energy is that some energy is converted to heat and is therefore not available to accomplish the task.

## Units of Work and Energy

Because work and energy are vital to our way of life, much jargon has developed to describe and measure it. To the uninitiated the units in which we measure things are a horrible maze. To find our way through the maze we can apply some simple ideas. Units should aid our understanding rather than add confusion. There are fundamental units from which all other units can be derived. For the sake of clarity we will consider only units in the SI or metric system.

The fundamental quantities are time, measured in seconds (s); length, measured in meters (m); mass, measured in kilograms (kg); and current, measured in amperes (A). Recall, from Newton's Second Law, that force is mass times acceleration and that acceleration is distance per time per time — distance/(time × time). Therefore, a reasonable unit for force would be a kilogram meter per second per second. Because this is more than one would want to say in polite conversation, we call a kilogram meter per second per second a newton (N). Work, as previously discussed, is force × distance, expressed in newton-meters or joules (J). Work and energy, because they are expressions of the same basic quantity, are both measured in joules. Power is the rate at which energy is converted (or work is done). The SI unit for power is the watt, which is equal to one joule per second. Therefore, a 60-watt lightbulb converts 60 J of electricity to light and heat every second.

## Quality of Energy

We mentioned that our interest is in useful energy. The more useful an energy is the higher its quality. We evaluate an energy's quality by how easily we can convert, transmit, and control it — the easier to convert, transmit, and control, the higher the quality.

Electrical energy is high quality. There is a well-developed technology to convert it to other useful forms. Motors convert it to mechanical energy, incandescent bulbs or fluorescent tubes convert it to light, and resistive elements to convert it to heat. Electricity is easy to control: It will flow through wires and not through insulation. Transmission over long distances is relatively simple.

Heat energy is hard to transmit and control. It flows readily from any warm object to any cool body. Insulation only slows the process. Except when used for space heating, heat usually is an intermediate step in an energy-conversion process. For example, burning coal (chemical energy) to produce steam to run a turbine uses heat energy as an intermediary between chemical energy and electrical energy. Heat is lower quality energy.

## *Efficiency*

We judge energy conversion by examining its efficiency. Generally, efficiency is defined by dividing the output energy by the input energy:

$$\text{Efficiency} = \text{energy out/energy in}$$

Should the energy out equal the energy in, the efficiency would be 1.00. In the real world the efficiency is never 1.00. If it were, perpetual motion devices would be commonplace. There always is some unintended conversion that produces energy in the wrong form. "Energy out" in the efficiency equation is defined as the useful energy.

Take, for instance, an electric motor. Its function is to convert electrical energy to mechanical energy; however, part of the electrical energy is converted to heat because of the resistance of the wires in the motor and part of the mechanical energy is converted to heat through the friction of the shaft rotating in bearings. The energy that is not recovered in a conversion is called the loss.

Energy transmission also has losses. When we send electricity through wires there is some resistance (think of it as electrical friction) that converts the electrical energy to heat.

## *Summary*

Energy is the capacity to do work. Work is the transfer of energy from one physical system to another. Energy appears in many forms. The methods of energy conversion, transmission, and control are called energy technology. When we examine energy usage we must consider both the efficiency of the conversion and the quality of the energy. We will try to bring these ideas to bear when considering specific technologies. More information about energy can be found in Ref. [1] or other standard engineering books.

## REFERENCES

1. Avallone, E. A., and Baumeister, T., III (Eds). (1989). *Marks Standard Handbook for Mechanical Engineers*, (9th ed.). McGraw-Hill, New York.

    *The standard reference, useful because it gives systems as well as physics and devices. Get it from any library.*

2. Krick, E. V. (1965). *An Introduction to Engineering and Engineering Design.* Wiley, New York.

*This readable, well-illustrated book is meant for first-year engineering students or someone deciding on engineering as a career. Nearly a third of the book describes the design process. Most of the rest describes what engineers do.*

3. Commoner, B. (1987, June 15). A reporter at large—The environment. *The New Yorker*, pp. 46–71.

   *Most of the environmental examples in this chapter are from this article. Commoner is worried and blames much of our trouble on the quest for profits.*

4. Riggs, J. L., and West, T. M. (1986). *Engineering Economics*. McGraw-Hill, New York.

   *The capital recovery factor is a standard technique and described in many places. This book is thorough and readable.*

## PROBLEMS

1. The design process described here can be used for many purposes. Choose an activity of some kind, such as a fund-raiser for an organization, and design the activity, going through each of the steps shown.

2. Evaluate the design of a system of interest to you, for example, airline passenger service or a football team's offense, along the lines of the test hardware and software questions discussed in this chapter.

3. List intangible factors relevant to the short-haul transportation systems mentioned in this chapter.

4. Consider who, in the long run, might benefit and who might suffer if each of the short-haul transportation systems mentioned in this chapter were implemented. Do the same people tend to benefit and suffer?

5. Compare two projects, both with lives of 15 years. One has an initial cost of $2500 and annual costs of $500. The other has an initial cost of $1200 and annual costs of $700. Assume the interest rate is 8%. Which project is cheaper? Would your conclusion change if the interest rate were 20%?

6. Why are projects with high initial costs generally riskier than ones with low initial costs, even if the sum of the annual payments over the life of the project are identical?

7. Table 2.1 shows that if the duration of a loan is doubled the capital recovery factor is not halved. Does this indicate that one pays more interest on a long-term loan than on a short-term one? Explain.

*Chapter*

# 3

· · · · · · · · · ·

# ELECTRICITY

· · · · · · · · · · · · · · · · · · ·

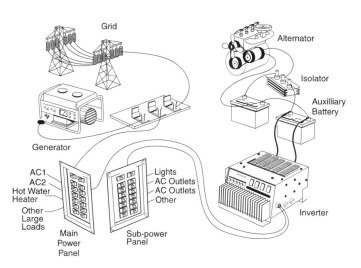

**Reproduced with permission from Real Goods Trading Corporation,**
***Alternative Energy Sourcebook***.

## NEED FOR POWER IN THE FORM OF ELECTRICITY

We study electricity because many of the appliances we want to use are powered only by electricity. Included among these are the ones that make life significantly better, such as radios, VCRs, and refrigerators. A home that most of us would want to live in must have an electrical supply. Electrical supplies, however, can be complicated and expensive and the difficulties and cost increase as the amount of power produced increases. It is probably not sensible to use electricity to cook a meal or warm a room when a simpler kind of energy would do. One of the purposes of this chapter is to learn enough about electricity to plan the capacity of the electrical supply. We will start with the basics.

## VOLTAGE, CURRENT, AND RESISTANCE

In this section, the fundamental aspects of electricity will be discussed. *Voltage* is analogous to pressure. It is the amount of force or push of the electricity. The units measuring voltage are *volts* (V). A flashlight battery pushes with 1.5 V and an automobile battery with 12 V (when new!). An electric outlet in the United States pushes with the equivalent of 120 V. The word "equivalent" is used because actually the voltage is changing continually, as we will discuss later.

*Current* is the amount of electricity that flows. Electric current actually consists of the motion of small charges, called electrons. Current is measured in *amperes* (A). An ampere is the flow of a certain number of electrons through a conductor in 1 s. Fuses or circuit breakers in a home will blow or trip when more than the rated current flows through them. A 15 A fuse will protect a circuit designed for 15 A by melting when more than 15 A flows through it. A 20-A fuse can carry more current without melting and so does not protect a circuit designed to carry 15 A.

A hydraulic analogy may be helpful. Consider a water tank on top of a hill (Fig. 3.1). The water pump shown refills the tank. A pipe leads from the tank to the load, where the water pressure is being used to do mechanical work.

The pressure of the water in the pipe is determined by the height of the hill. Voltage is analogous to this pressure. Current is analogous to the amount of water that flows through the pipe in gallons per minute or,

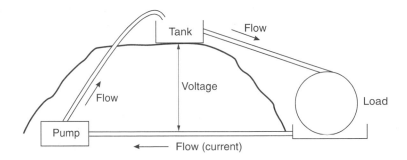

**Fig. 3.1. Hydraulic analogy for electricity.**

slightly more accurately, drops per second. Note in Fig. 3.1 that a return path for the water is shown. Electric current always flows in a complete loop or circuit. In some cases the return path can be through the ground or the metal frame of the structure, as in an automobile, but usually the return path is a wire.

The amount of water that flows depends, of course, on the pressure or the height of the hill. It also depends on the width of the pipe carrying the water—the narrower the pipe, the less water. The electric analogy to pipe size is *resistance*—the greater the resistance the smaller the current. Resistance is measured in ohms ($\Omega$). A 4-$\Omega$ loudspeaker will allow twice as much current to flow as an 8-$\Omega$ speaker, which is why it is risky to replace an 8-$\Omega$ speaker with a 4-$\Omega$ one. (Amplifiers can be damaged if too much current is drawn from them.)

Current, voltage, and resistance are related as shown in Eq. [3.1], which states that current increases as voltage (or pressure) increases but decreases as resistance increases. *I* is current, *V* is voltage, and *R* is resistance:

$$I = \frac{V}{R} \qquad [3.1]$$

Often Eq. [3.1] is written as Eq. [3.2]. This relationship is known as Ohm's Law:

$$V = IR \qquad [3.2]$$

### *Example of Ohm's Law*

Commercial electric wire has various thicknesses: The thicker the wire, the more cross-sectional area for the current to flow through so the smaller the resistance. (It takes less pressure to push water through a large-diameter pipe compared to a small one.) A so-called No. 4 wire 100 ft long has a resistance of 0.025 $\Omega$, whereas the same length of No. 10 wire, which is thinner, has a resistance of 0.1 $\Omega$. If 80 A flows through the No. 4 wire then the voltage lost in the wire, that is, the pressure used up pushing the current through the wire, is 2 V:

$$V = 80 \times 0.025 = 2 \text{ V} \qquad [3.3]$$

The voltage loss in a No. 10 wire (resistance of 0.1 $\Omega$) would be four times as large:

$$V = 80 \times 0.1 = 8 \text{ V} \qquad [3.4]$$

One reason the lower resistance wire is not always used is cost. One 100 ft length of insulated No. 4 wire would cost about $120, whereas 100 ft of insulated No. 10 wire would cost about $23.

## POWER

A fourth quantity exists besides voltage, current, and resistance: *power*. Power is the amount of work done per second. Power is measured in *watts* (W). A 100-W amplifier puts out twice as much power as a 50-W amplifier. By law, an appliance manufacturer must indicate on an appliance how much power the appliance uses. (In some cases current rather than power needs are given.) A typical toaster, for example, uses 1200 W, whereas a vacuum cleaner uses 520 W. An electric company charges for the total amount of work done, or energy used, so an electric bill shows charges for "watt-hours" or, more usually, "kilowatt-hours" (A kilowatt is 1000 W). Power is the product of voltage and current:

$$P = V \times I \qquad\qquad [3.5]$$

This equation makes sense. One can increase work done or power consumed by pushing electrons harder or by pushing more of them; in other words, by increasing voltage or by increasing current. The hydraulic analogy may help here. The amount of water power available to the load in Fig. 3.1 is increased by increasing the height of the tank or by letting more water flow through the pipe.

The formula for power (Eq. [3.5]), can be used to calculate the amount of current an appliance uses. The reason one cares about the amount of current that will flow is that not caring may result in a blown fuse. Appliances sold in the United States are generally meant to be used with 120 V. If an appliance is rated at 1200 W, it uses 10 A from a 120-V circuit:

$$P = V \times I$$
$$1200 = 120 \times I \qquad\qquad [3.6]$$
$$\text{or } I = 10 \text{ A}$$

If a home has 15-A fuses, then the total current in one circuit must be less than 15 A so other appliances plugged into the same circuit as this 10 A appliance must use less than 5 A. A 60-W lightbulb uses 0.5 A, so the number of 60-W bulbs that can be powered from a 15-A circuit is 30. Table 3.1 provides a summary of the fundamental electric quantities.

## Table 3.1. Fundamental Electric Quantities

| Quantity | Description | Unit | Abbreviation | Symbol |
|---|---|---|---|---|
| Voltage | Pressure | Volt | V | V |
| Current | Amount of flow | Ampere | A | I |
| Resistance | Resistance | Ohm | Ω | R |
| Power | Voltage × current | Watt | W | P |

# HOW MUCH ELECTRIC POWER DOES A FAMILY NEED?

In order to design an electric system, one must estimate needs. The power needs of some appliances are given in Table 3.2.

The stereo, which uses 100 W probably only puts out 50 W of sound—the rest of the input power is given off as heat. In estimating electric power needs one must consider the input power required, not the desired output power. The other remark concerns the refrigerator. The power usage shown applies when the motor is on. Most of the time the refrigerator motor is off because only when the temperature rises above the thermostat setting does the motor turn on. The refrigerator takes power, therefore, only about a quarter of the time it is plugged into an electric outlet. The length of time it is actually drawing power depends on how well it is insulated and how often the door is opened. Also, most motors, including refrigerator ones, draw much more current when starting compared when running—just as a car requires more gas when accelerating than when going at a constant speed. The starting current, which lasts for only a short time, is about four times the rated current. Because the starting current is large, time delay fuses are used in homes. These fuses will only blow if the current exceeds the rated value for 5 s or so—long enough to start the motor but not long enough to overheat the house wiring. It takes time for the house wires to get hot enough to do damage, just as it takes time for a toaster to get hot.

Using Table 3.2, we estimate minimal electric energy needs for a small family. The question is not so much what we would like but what is really needed for a reasonable life style. Of course, each person's conception of what is reasonable differs, which is why technology planning is difficult. We might start by deciding that we need power from about 6:00 to 11:00 PM, that is, 5 hours. During this time we need four 75-W lightbulbs on at once plus a stereo. We might also agree that either a sewing machine or an

### Table 3.2. Power Needs of Appliances

| Appliance | Power needs (W) |
|---|---|
| Transistor radio | 4 |
| Small stereo | 100 |
| Black-and-White TV | 45 |
| 3/8-in. Portable drill | 260 |
| Refrigerator | 300 |
| Sewing machine | 100 |
| Water pump | 335 |

electric drill would be used, but not both at once, again for 5 hours. We certainly will need a refrigerator, but if we are careful it will draw power less than 5 hours a day. If we need a water pump, then again we have to estimate how many hours it will be on. Can it fill a tank when no other appliances are on? For now, let us ignore the water pump. Our minimal power needs are shown in Table 3.3.

Our daily energy needs are 960 W for 5 hours or 4800 Watt-hours. Thus, we must design our system to supply 960 W at any one time and to supply 4800 watt-hours (4.8 kilowatt-hours) over a day. It is worth noting how much of the power is going into incandescent light bulbs. In our design we really should have specified fluorescent bulbs, which use much less power to produce the same amount of light but cost more. In the following sections we will discuss how to design the system.

### Table 3.3. Minimal Electric Needs

| Appliance | W | hr |
|---|---|---|
| 4 75-W lightbulbs | 300 | 5 |
| Stereo | 100 | 5 |
| Sewing machine or electric drill | 260 | 1 |
| Refrigerator | 300 | 5 |
| Total | 960 | 5 (worst case) |

## PRACTICAL GENERATORS

The principle of electric power generation is simple. The basic physical principle is that a changing magnetic field through a coil of wire produces a voltage across the ends of that coil. The construction of a generator is straightforward. A generator is shown in Fig. 3.2. A magnet, usually an electromagnet, is rotated inside a stationary coil of wire. In Fig. 3.2 the stationary coil is really a set of coils connected together. These are labeled "stator" and the rotating electromagnet is labeled "rotor." A small current is fed into the rotor through the slip rings, producing a magnetic field that points in a direction perpendicular to the rotor shaft. As the rotor rotates inside the stator the magnetic field in the stator coils changes. This changing magnetic field produces voltage in the coils. The coils are connected in series and the two ends lead to the electric load, that is, these ends of the coil go to the generator terminals. An electric company's generator terminals are connected to transformers, switchgear, and trans-mission lines but ultimately go to the metal pieces inside the electric outlet in a home. When an appliance is plugged into the outlet the circuit is completed — current flows out of one end of the coil, through the connect-ing wires to the load, and back from the load to the other end of the stator coil. The voltage produced by the rotating electromagnet is the force that drives the current through the circuit.

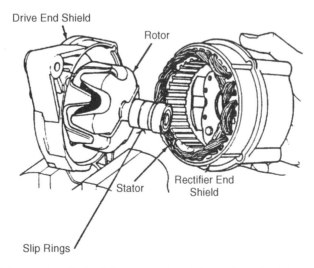

**Fig. 3.2. Schematic of electric generator. (Reproduced with permission from Chrysler Corporation.)**

The circuit is completed slightly differently in an automobile and some other applications. In an automobile one end of the generator coil is attached to the metal frame of the car and the other to the load. The load must then be attached between the wire from the generator and metal frame. Because only one wire is used it appears as if no return path for the current exists. Current has to flow, however, in a complete circuit — from the generator to the load and back — because the generator does not produce electrons. It only pushes them along the circuit.

While the principle of a generator is simple, construction is not easy. Because the generator turns rapidly, balance and good bearings are necessary. Also, winding coils is tedious. The result is that it does not make sense to try and build one's own electric generator, unless one has access to a machine shop, a coil winder, and also has some experience.

Every automobile contains a generator like the one shown in Fig. 3.2, which recharges the battery and powers the lights, spark plugs, radio, and so on. One possibility, then, in a homemade electric power system is to use scrap automobile components. Another possibility is to buy a commercial generator.

In an automobile a device called a "voltage regulator" is placed between the generator and the battery. Voltage regulators are needed because the voltage from the generator depends on the engine speed, which varies. The magnitude of the voltage from a generator depends on how fast the magnetic field is changing; when the generator is rotating rapidly, the voltage produced will be large, and when the generator is going slowly, the voltage out will be small. The generator is driven by the car's engine so when the automobile is going fast, as on an interstate, the battery is charged rapidly. When the engine is idling at a light the voltage produced is small. Why do we care about the variation in voltage? One reason is that a high voltage might damage the battery. Another is that when the generator voltage is less than the battery voltage, current will flow back from the battery into the generator. The voltage regulator prevents this battery discharge at low generator speeds and also reduces the output voltage at high speeds. A regulator should be used in a home-built system, although an operator could monitor the output and make adjustments manually. The voltage regulator is built into most modern automobile alternators, so it is hard to avoid getting one if an alternator for a recent car is purchased. A commercial electric generator will usually have a built-in voltage regulator and, as one would expect, the utility companies have regulator mechanisms for their generators.

Because spare auto parts are so easily available, it is tempting to design a home-built system around them. Before we settle on a 12-V system, however, we discuss the implications of some of the calculations we did earlier. If we need 960 W and will use 12 V then, according to Eq. [3.5], we would need 80 Amperes:

$$960 = 12 \times I$$
$$\text{or } I = 80 \text{ A} \tag{3.7}$$

If we want to live some distance from the generator to reduce the noise, for example, we would have to run the electric current through connecting wires. If our house was 50 ft from the generator, then we would need 100 ft of wire to get a complete circuit. We determined from Eqs. [3.3] and [3.4] that if we used No. 4 wire our voltage loss along the 100 ft of wire would be 2 V so of the 12 V generated only 10 V would be available to do anything useful. If we tried to save money and used No. 10 wire, which has a higher resistance, we would lose an intolerable 8 V. Actually, No. 4 wire is very thick and difficult to handle, and even No. 10 is thicker than that used in most homes. Also, 20 ft of wire is probably used up internally just going from outside of the house to the outlets, so with 100 ft of wire, the house and the generator could be only 30 ft apart.

It appears that a 12-V system has a real problem; to get sufficient power in the load, one needs large currents, which mean large voltage losses in the wiring. The usual solution is to use higher voltages, which is why the standard voltage in U.S. homes is 120 V, and in most of the rest of the world it is 220 V. Small 120-V generators are not nearly as common as automobile generators and are much more expensive. The catalog described in Ref. [3] has small electric generators.

## DC and AC Power

Another difference exists between the voltage used in an automobile and that in a home — DC vs AC (Fig. 3.3). The abbreviation "DC" stands for direct current, meaning that current always flows in the same direction — one generator terminal is always positive and the other is always negative. Batteries give DC voltage and must be recharged using DC. In fact, the reason DC is used in automobiles is just that — to recharge the battery. The reason 12 V are used is that the chemical process inside a single battery cell produces only 2 V. To get higher voltages, cells must be connected in series, and six cells are as many as are practical; otherwise, the battery would be too big to fit into the automobile conveniently. Also, the connections between cells are costly, so the manufacturer does not want to use more cells than necessary.

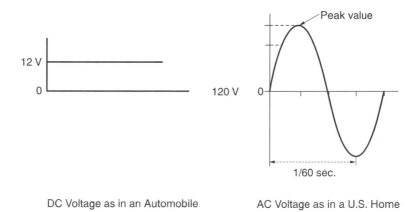

DC Voltage as in an Automobile          AC Voltage as in a U.S. Home

**Fig. 3.3. DC and AC voltage.**

AC voltage is actually much more common than DC. One reason is that the generator in Fig. 3.2 and most generators actually produce AC, as one might expect from a rotating device. As the magnet rotates the magnetic field changes direction. Other devices, called "rectifiers," are used to convert the AC to DC. In most present-day automobiles the generator is actually called an "alternator." The rectifiers are an integral part of the alternator so that the voltage into the battery is DC. Rectifiers are discussed in the Appendix to Chapter 7.

A major advantage of AC is that transformers can be used to change the voltage level. Step-up transformers increase the voltage but decrease the current; the power is not changed from the input to the output of a transformer. Step-down transformers decrease the voltage and increase the current. Utilities generate at medium voltage, transform to high voltage for transmission, and transform down to low voltage for the user. High-voltage transmission means low current and thus small losses in the lines. Low voltages, though, are safer, which is why the user is given 120 V. The reader may have noticed large, metal, cylindrical-shaped objects on electric utility poles. These are transformers, which step down the voltage from 440 V, used for local transmission, to household voltage.

A final advantage of AC is that it can drive motors at a constant speed, corresponding to the number of times the voltage goes up and down in 1 s. In the United States the voltage alternates 60 times a second. The number of alternations per second is called frequency. In most of the rest of the world the power company frequency is 50 cycles per second. The reason we care about the frequency is that we must make sure it is correct, otherwise electric clocks or record-player turntables will not rotate at the proper speed.

As mentioned before, rectifiers convert AC to DC. Devices called "inverters" convert DC to AC. Inverters are more complicated than rectifiers, partly because they must produce the correct frequency. A typical price for a 1000-W inverter is about $450 [3]. Inverters also waste some power, so only about 80% of the DC input power is converted to AC output power—the rest is given off as heat. Inverters are basically switches which turn the DC voltage on and off. We would use an inverter if we had a DC power source, such as a battery, and wanted AC power.

## STORAGE

Some generators, such as those powered by wind or solar, do not produce power continuously, so the electricity must be stored. The easiest way to store electricity is with batteries but, as mentioned before, only DC can be stored in batteries. Other methods of storage are cumbersome but possible. Utilities sometimes store electricity by pumping water into a reservoir on top of a mountain when excess power is available and then using that water to drive turbines when power is needed. In this way, the peak power from a steam-powered generator can be greater than the capacity of the generator. Because both the pumps and turbines are AC equipment, AC electricity is being stored.

A scheme analogous to storage is to sell excess electricity back to the utility company. This is straightforward in practice since when current flows through the meter in the opposite direction it drives the meter backwards, reducing the bill. When one generates more electricity than one is using then electricity is being sold to the utility company. When more electricity is being used than is being generated, electricity is being bought from the utility. This process of buying and selling amounts to storing electricity with the utility. By law a utility must buy electric power from homeowners, although the price paid may be different from the price charged. The utility insists that the frequency of electricity sent in be exactly right so the generator or regulator in the homeowner's system must be accurate. Of course, such storage with the utility is not possible if one is not connected to the power grid.

Batteries are a relatively mature technology: in other words, most of the complications have already been worked out. Automobiles use so-called lead-acid batteries; the plates are lead and the fluid is sulfuric acid. When charging, a chemical reaction takes place at the plates. At discharge the reverse reaction takes place, letting current flow out. Nickel-cadmium batteries are similar, somewhat smaller, and more reliable but more expensive. Lead-acid batteries are probably the most practical for a home-built electrical system because they are readily available. Automobile batteries actually are designed to give a large amount of current for a few seconds to start the engine, which is a different application than giving sustained current for several hours or even days, so in a home system a so-called "deep cycle" battery, designed for golf carts, recreational vehicles, or small boats, is used. A typical deep-cycle battery might supply 9 A for 10 hr or 15 A for 5 hr before needing recharging. Such a battery costs about $80. Batteries do need care, which consists mostly of checking the fluid level. They work best when warm and need protection from freezing.

## OVERALL SYSTEM

Now we will design the overall system. We have several choices. We will opt for a simple one that could be built from scrap automobile parts. The system is shown in Fig. 3.4.

We review what is in each block. The "mechanical driver" turns the generator. It could be a waterwheel, a wind-driven propeller, or a diesel motor. The automobile generator produces 12 V DC. The generator is probably really an alternator with built-in rectifiers. Such alternators/rectifiers are available at auto supply stores or junkyards. Because so many are available, the price is low (perhaps $50), although the operating life may be only 3 or 4 years. The voltage regulator shown is the same as used in automobiles and is purchased from the same places. It controls the amplitude of the voltage produced as the generator speed varies. The batteries store energy during the times when the generator is not moving. Deep-cycle batteries should be used because the batteries must produce current for approximately 5 hr. The number of batteries to be used is calculated in the following section. The fuses protect the system if something goes wrong and a harmful amount of current would be drawn from the batteries. The "DC Electric Loads" are special appliances which use 12 V DC. Many of these are made to be used with recreational vehicles. The

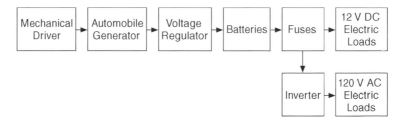

**Fig. 3.4. Economical home-built electric system.**

inverter converts 12 V DC to 120 V AC so ordinary household appliances can be used. An advantage of 120 V over 12 V is that the same power can be transmitted at a lower current, which means less loss in the connecting lines. Actually, one would probably not use both AC and DC, especially since the inverter is one of the most expensive parts of the system. Here though, we are assuming we need some of our output power as AC.

## COST OF THE SYSTEM

Representative prices for each component are shown in Table 3.4. A reasonable assumption is that each of these would last 3 years. (Actually, the fuses and connections should last nearly indefinitely.)

If we assume the capital recovery factor is 0.41 (3 years, 12%), the annual cost is

$$0.41 \times 855 = \$350/\text{year} \qquad [3.8]$$

The reader might compare this amount to her or his annual electric bill. Note that the costs associated with the mechanical driver (waterwheel or windmill) have been omitted.

One might decide to use 12-V DC loads only and spare the cost of the inverter. While most appliances are designed for 120 V one can buy refrigerators, vacuum cleaners, and other appliances for recreational vehicles which use 12 V DC. Of course, automobile lights and radios use 12 V DC. A 12-V refrigerator costs \$520, a vacuum cleaner costs \$25, and a fan costs \$24. Use of a 12-V system does mean one must be wary of losing significant voltage along the connecting wires, so the generator must be close to the load.

### Table 3.4 Prices of Components of Electric System

| Component | $ |
|---|---|
| Used automobile generator | 50 |
| Used voltage regulator | 15 |
| 6 deep-cycle batteries | 480 |
| Fuses and connections | 10 |
| Inverter (500 W) | 300 |
| Total | 855 |

An alternative to the homemade system of Fig. 3.4 would be to buy a commercial AC generator. These are sold for various applications and some are discussed in Chapter 5. Others are meant to be used with diesel engines for emergency power. They cost a good deal more than our system (approximately $3000) but would probably last longer.

## FINAL WORD

One purpose of this chapter was to estimate how much electric power is needed in most people's lives. Another purpose was to show that if a source of mechanical energy is available, it is not difficult to design and build a simple system meeting basic needs. The process of doing one's own design gives a much better appreciation for how the system works than reading a catalog or listening to a salesperson and, in any case, helps greatly in evaluating what one has read or heard. Listening to experts can be dangerous unless one has background to evaluate the advice.

## REFERENCES

1. Leckie, J., Masters, G., Whitehouse, H., and Young, L. (1981). *More Other Homes and Garbage*. Sierra Club Books, San Francisco.

   *Contains a wealth of detailed information. The material in this chapter also appears in various places in their section "Small Scale Generation of Electricity from Renewable Energy Sources." They give more detail and start at the elementary level so anyone can benefit.*

2. New Alchemy Institute, *The Journal of the New Alchemists*, Woods Hole, Falmouth, MA 02543.

   *Volume 2 has a good summary of various electrical system with a focus on wind generators. See the article by Frederick Archibald, "Windmill Electronics." Explains AC and DC and gives information about automobile alternators and batteries.*

3. Real Goods, *Alternative Energy Sourcebook 1991*, Real Goods, 966 Mazzoni Street, Ukiah, CA 95482 ($10).

   *The subtitle is true — "A Comprehensive Catalog of the Finest Low-Voltage Technologies." Photovoltaics to waterless composting toilets, with some explanatory articles on electrical systems.*

4. Salm, W. (1965). *Home Electrical Repairs Handbook*, No. 736. Fawcett, Greenwich, CT 06830.

   *Bookstores often have guides for people wiring their own homes or repairing their own appliances. This one has much practical information on wire losses, the electrical code, and house wiring.*

*Catalogues for automobile supply companies and home building companies have current prices on auto alternators, etc. and wire.*

## PROBLEMS AND PROJECTS

1. How much voltage would be lost if 8 A went through 100 ft of No. 4 wire? Through No. 10 wire?

2. Repeat Problem 1 if the wire length is 200 ft.

3. Choose several electric devices important in your life and estimate the total current required when all are being used — assume the voltage is 120 V. Probably not all would be used at once, so estimate the total amount of current that would be used at any one time.

4. Continue Problem 3 by estimating the number of kilowatt hours these appliances will use in a day. If your utility charges 10¢ a kilowatt hour, what would the daily bill be? The typical monthly bill?

5. Do the calculations of Table 3.3 for a family following your lifestyle make sense?

6. Determine the costs of a new alternator, a rebuilt alternator, and a deep-cycle battery in your locality. Confirm if the rectifiers and voltage regulators are built in, and if not, obtain a price for these also.

7. As we will see later in the book, and as was already mentioned, electric power is usually more complicated to generate than heat or mechanical power (solar energy gives heat directly, whereas water or wind give mechanical power). Consider how power is used in your house, or perhaps in a vacation home, and decide which needs must be met by electricity and which can be met in other ways.

8. Assume we want to transmit 960 W from a generator to our house, using 100 ft of wire. Compare the voltage lost in the connecting wires (No. 4 gauge) when 120 V is used to that when 12 V is used. (This problem is based on Problem 1.)

9. Assume we wanted 480 W continuously for 10 hr. How much current is needed from a 12-V system? How many batteries of the type describe in this chapter are needed? If the generator recharging the battery can supply 360 W how long must it be on to do this recharging? (Hint: The energy in watt-hours available from a battery equals the energy put into that battery.)

10. Price out the system of Fig. 3.4 if all the electric appliances could use 12-V DC. Estimate the monthly cost if the money were borrowed at 10% for 5 years.

11. An alternative to an automobile alternator, which will probably have to be replaced within 5 years (and maybe sooner), is a commercial alternator costing perhaps $3000. Describe the advantages to you of the cheaper system and those of the more reliable system.

12. (This project requires actual construction.) Design and build a bicycle-powered system that generates electricity. Use a lightbulb as the load. A scrap automobile generator might be used.

*Chapter*

## 4

.............

# HYDROPOWER

.....................

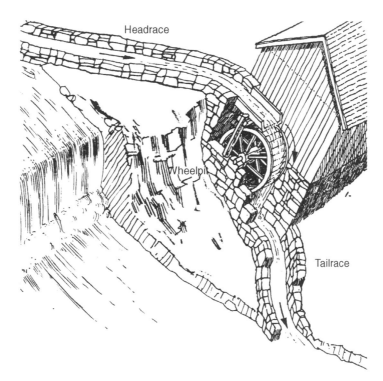

**Reproduced with permission from Houghton Mifflin Company,
David Macauley, *The Mill*.**

## APPALACHIAN MOUNTAIN CLUB CABIN

The Appalachian Mountain Club maintains a cabin for hikers near Twin Mountain in Vermont. The power needs of the cabin are a water heater, a refrigerator for medical supplies, a food and beverage warmer, and a two-way radio set. In the past, the radio was powered by batteries and the rest by butane. The butane and fresh batteries were brought in by helicopters, which are expensive and noisy and inconsistent with the wilderness environment. Also, helicopters could not reach the cabin during bad weather. Bringing in power lines would be much too expensive. A stream flows by the cabin with a 202-ft drop. The obvious solution — water power — worked. A hydroelectric generator powered by a water turbine was installed and replaced the butane system. The turbine–generator system is almost maintenance free. It will, over the long run, cost less than butane and batteries and will be more reliable.

This real situation is one in which small-scale hydroelectricity is the best solution. Other situations exist in which it is cheaper to generate one's own electricity by harnessing a stream than by bringing in power from the electric utility. In rural New England, for example, the cost of installing electric lines is about $20,000 a mile.

## AMOUNT OF POWER AVAILABLE

The energy driving a waterwheel comes from the force of the falling water. The amount of potential energy depends on how far the water can drop: the greater the drop the greater the energy. Of course, the amount of energy available in falling water depends also on how much water is falling. Power, as noted in Chapter 2, is the energy available per second, so power depends on both the height of the drop and the quantity of water dropping per second.

The equation for the power which can be generated from a stream is as follows:

$$P = \frac{HQe}{11.8} \qquad [4.1]$$

where $H$ is the drop in feet, $Q$ is the amount of water in cubic feet per second (ft$^3$/s), $e$ is the efficiency of wheel and generator, and $P$ is the output power in kilowatts.

This formula makes sense. It shows that the power increases when distance dropped increases. (Incidentally, this drop is usually called the "Head.") It also shows that the more water, $Q$, that drops, the more power is available. Both the wheel and the generator waste some power, so the actual power obtained is less than the power available. This wastage is accounted for by the efficiency, $e$, which is <1. The factor 11.8 is used to convert units, so the power comes out in kilowatts, when $H$ and $Q$ are measured using feet and cubic feet per second.

A picture of an overshot wheel, probably the easiest to build, is given in Fig. 4.1. The principle of operation is probably clear. The water flows from the "flume" into the buckets. The weight of the water turns the wheel. The "sluice gate" controls the amount of water that flows. The head, $H$, is measured from where the water leaves the flume to where it falls out of the buckets. Because some of the water falls out early, and because energy is lost in bearing friction, the efficiency of the wheel is about 75%. In most cases a waterwheel would drive a generator through belts or gears because the waterwheel itself rotates much slower than the rated speeds of most generators (think of how fast an automobile generator rotates). These belts or gears use up energy and their efficiency is also about 75%, so the overall efficiency of the wheel–turbine system is between 50 and 60%.

Now we calculate the power available from a small stream in terms of water velocity and steam size. First we will see how we would calculate $Q$, the quantity of water which flows through our wheel in a second. A sketch of the flume with an opened sluice gate is given in Fig. 4.2. The cross-sectional area of the flume is $A$ square feet. The velocity of the water is $v$ feet/second. The amount of water which flows past the opened sluice gate in 1 s is all the water which was within a distance $v$ ($\times 1$ s) from the sluice, because water closer will have time to get to the gate and water further away will not. So the amount of water reaching the wheel in a second is

$$Q = A \times v \qquad [4.2]$$

Using Eq. [4.1], the following is derived:

$$P = \frac{HAve}{11.8} \qquad [4.3]$$

Now we can estimate the power from a typical waterwheel. Assume the head is 9 ft, corresponding to the flume being about 10 ft above the tailwater, because clearance is needed above and below the wheel. Assume the cross-sectional area of the flume is $2.0 \times 0.66$ ft, that is,

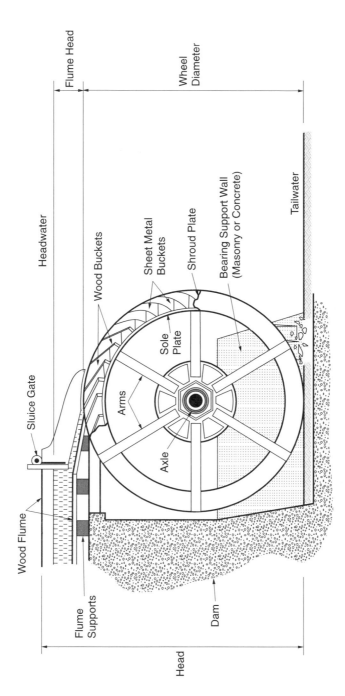

**Fig. 4.1. Overshot waterwheel. (Reproduced from Mother Earth News [3].)**

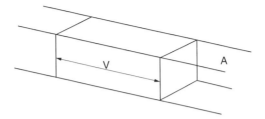

**Fig. 4.2. Water flowing in flume.**

24 × 8 in. Assume the water is moving at 2 ft per second. We will estimate the overall efficiency at 55%. The power is calculated as

$$P = \frac{9 \times 2 \times 0.66 \times 2 \times 0.55}{11.8} = 1.13 \text{ kW} \qquad [4.4]$$

This amount of power is more than enough to meet the needs of the family of four discussed in Chapter 3. Note that the flow, $Q$, is 2.64 ft$^3$/s, as shown in Eq. [4.5]:

$$Q = AV = 2 \times 0.66 \times 2 = 2.64 \text{ ft}^3/\text{s} \qquad [4.5]$$

In other words, a stream carrying 2.64 ft$^3$/s and dropping 10 ft would have sufficient power to meet the minimal needs of a family.

## SYSTEM DESIGN

Now that we understand the basic physics of water power, we examine how we would design a whole hydroelectric system. A block diagram in Fig. 4.3 shows the steps required in designing a small hydroelectric system for a given stream. We will discuss these steps separately.

### Verifying That Sufficient Power Exists

If we are to build a hydroelectric system, we must certainly have a stream flowing through our property. We need to estimate both the head, $H$, and the quantity of water, $Q$, to use Eq. [4.1]. The head is the distance the water drops within the length of stream to be used. It can be estimated using a string and tape measure (Fig. 4.4). To do this one would tie one end of the string to a stake driven close to the water level, move downstream as far as practical, pull the string taut and horizontal (a line level would aid in getting the string horizontal), and measure the vertical distance from the

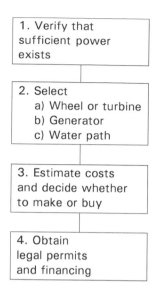

Fig. 4.3. Steps in system design.

string to the water level. One would repeat this process until the useful length of the stream is covered. Another possibility is to use a surveyor's level.

The quantity of water in a flume, of course, cannot be greater than the quantity of water in the stream. To estimate this we need the cross-sectional area of the stream and the velocity. Estimating the cross-sectional area is basically simple but water flows more slowly near the banks and the bottom; therefore, in estimating the area one disregards the 3 in. or so near the banks and bottom. An easy way to estimate the velocity is to put a twig, or something else which floats, in the water and time it as it goes a fixed distance. More sophisticated ways of measuring the stream velocity are given in the books *More Other Homes and Garbage* [2] and *Mother's Energy Efficiency Book* [3].

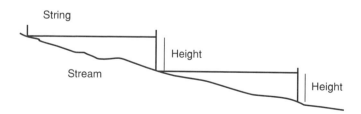

Fig. 4.4. Side view of a stream.

Water flow is seasonal, so if one is serious about using the power, one should examine the stream at different times of the year. In New England, August tends to be the low-water month. Streams will usually flow under ice in the winter, so winter operation is possible. Actually the most dangerous time for small hydropowered installations is probably in the spring, when the whole mechanism could be lost in a flood. It is prudent to check the stream at each of these times — spring, winter, and fall. It is also prudent to be wary of even careful estimates, especially if they are close to the limits of what is needed or can be handled. Besides the errors in making the estimates, one must realize that nature is not dependable. If one needs 960 W and the estimate of the output power is only 1200 W it is probably best to redesign; 1200 W is really too close to 960 W to be reliable. One should increase the power, if possible, or decrease the load required.

Once the head and quantity of water in the stream are known, then we can use Eq. [4.1] to calculate the maximum power available from the stream. The next step is to select a waterwheel or turbine.

## Types of Waterwheels and Turbines

The expression "waterwheel" usually refers to a large-diameter (approximately 10 ft) wooden wheel which turns relatively slowly, perhaps one revolution per minute. The word "turbine," on the other hand, refers to a smaller metal wheel, approximately 2 ft in diameter, which turns much faster — several thousand revolutions per minute. Waterwheels are probably easier to make, although they are large and heavy and therefore difficult to move from the shop to the site. Detailed plans are given in the books in the References. To make a turbine, one really needs access to a machine shop. Again, detailed plans are given in Refs. [1] and [2].

Three types of waterwheels are shown in Fig. 4.5. As mentioned previously, the overshot one is the easiest to build and is the most efficient (65–85%). It needs a head of at least 10 ft. In principle, one could use as large a head as available, but because the diameter of the wheel and the head are approximately equal, heads over 30 ft are not feasible for an overshot wheel.

The breast wheel is used for heads between 6 and 10 ft. The wheel shown is a low breast wheel because the water enters below the center of the wheel. Such a wheel has an efficiency of 65%. The problem with such wheels is the breastwork, the close-fitting wooden cover that keeps the water in the buckets. These need to be made carefully to reduce friction. Debris floats down the river and can get caught between the wheel and the breastwork so a screen is needed. Breast wheels can be made with large diameters, independent of the head, and large diameters create high

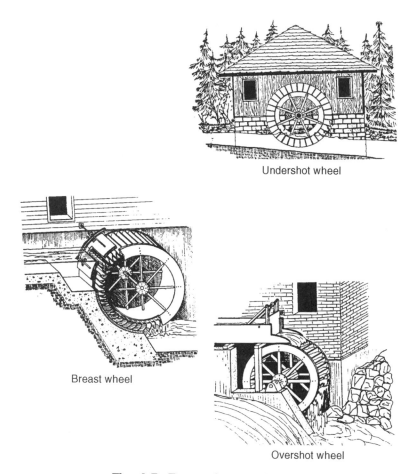

Undershot wheel

Breast wheel

Overshot wheel

**Fig. 4.5. Types of waterwheels.**

torque. High torque is useful if the wheel powers machinery or tools or grinding stones but is not important for electric generators, so breast wheels are not used often for hydroelectric projects.

Breast wheels work better than overshot wheels in flood conditions because the wheel turns in a direction which pushes the spent water away from the wheel. In an overshot wheel, in flood conditions, the wheel pushes water into the space behind the wheel. This water acts as a drag on the wheel. Breast wheels were used more often than overshoot wheels in colonial New England textile mills because of this ability to get rid of the used water.

The undershot wheel works with the lowest head. It is also the least efficient (25–45%). Here, the kinetic energy of the water pushing on the blades creates the force.

### *Turbines*

Two turbines are shown in Fig. 4.6. Turbines are the most efficient way to generate electricity. Commercial systems all use turbines. Many small-scale systems also use turbines. The efficiencies range from 75 to 85%.

The Pelton wheel uses a nozzle to direct the water onto the blades at a high force. These turbines are used in high head situations, especially when the total amount of flow is small. A small commercial turbine and generator producing 1.4 kW costs about $1400 [2]. Other commercial turbines are described in Ref. [6].

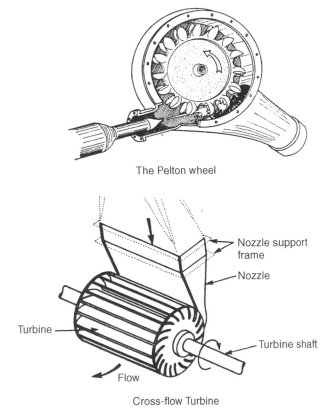

The Pelton wheel

Cross-flow Turbine

**Fig. 4.6. Cross-flow turbine. (Reproduced with permission from Mother Earth News [3].)**

A cross-flow turbine works on another principle. The water flows through the blades on the top, pushing them, then through the central cylinder and, finally, out through the bottom, pushing on those blades also. The word "push" is misleading. In Chapter 5 we will discuss "lift" and "drag" devices — the crossflow turbine actually uses lift. A careful amateur can make this type of turbine in a machine shop. Mother Earth Plans sell drawings [4]. If these plans are used, the cost will be about $50. A cross-flow turbine will work with heads as low as 3 ft.

A final type of turbine, the reaction-type turbine or Francis turbine, will not be considered further because they are only used in bigger installations. They usually are totally immersed in the water, with the rotating axis vertical. The weight of the water pushes on the blades. Such turbines were commonly used in the large textile mills built along rivers in the late nineteenth century, such as in Lowell, Massachusetts. Incidentally, the extensive waterworks at Lowell have been restored and are well worth a visit.

## Waterworks

In planning channels for bringing water to the wheel and back to the stream (step 2c in Fig. 4.3), a major decision is whether a dam will be built. Dams are needed if a pond is required to store water. If one needs electricity for only 6 hours a day, then one could store water for the other 18 hours. In this case the power available when used is four times the continuously available power. Actually, ponds are probably used most often to store energy over longer periods, such as during the summer, so the spring flood water is available in the dry months, such as August. As will be shown later, a dam can also reduce the length of the channel used to bring water to the wheel.

On the other hand, a dam is not a trivial thing to build, and a dam failure can be a major disaster for people downstream. The process of obtaining permits certainly will be easier if a dam is not used, so advantages exist in simply diverting water from a stream and returning it.

Figure 4.7 shows a side view and a top view of a hydroelectric site. The head race is the channel dug out to carry the water to the flume, and the tail race transports the water away from the wheel. In order to get the water to flow with sufficient speed, these must have a downward slope, perhaps 2:1000, which means a drop of 2 ft in a 1000-ft channel. If the water does not flow rapidly, a wider race must be used. Races reduce the head, so one has to trade off the additional drop in the stream gained from longer races with the loss in head in the channel. In colonial America some races were a quarter mile long. The total head available includes the height

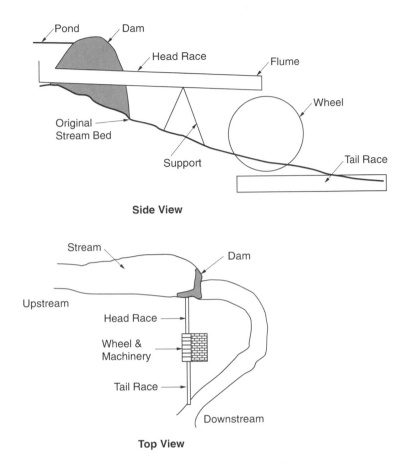

**Fig. 4.7. Hydroelectric site.**

of the water in the pond above the outlet as well as the drop through the wheel, but the usable head is measured from where the water drops on the wheel.

If a turbine were used in the same installation as Fig. 4.7, it would be located at the lowest point, where the water goes back into the stream. Instead of using an open channel for the head race, a polyvinylchloride (PVC) or metal pipe is used to enclose the water going from the stream to the turbine. This forces the water to flow through the nozzle in a Pelton wheel, or directly into the cross-flow wheel. This pipe is called a "penstock." PVC pipes of a suitable size (approximately 4 in.) cost about $0.50 per foot.

## *Size of Pond*

We now estimate how large a pond must be to store sufficient water to provide power for a week. In Eq. [4.4] it was shown that if the water flow is 2.64 ft$^3$/s the power will be 1.13 kW, sufficient for our needs. To simplify the analysis we will assume no water flows into the pond while we are drawing water out and that we draw water out 24 hr a day for the entire week. How much water must be stored to allow 2.64 ft$^3$/s for 1 week? We convert seconds to minutes to hours to days to weeks and multiply by 2.64 (Eq. [4.6]). We need a pond of about 1.6 million ft$^3$. How big is that?

$$60 \times 60 \times 24 \times 7 \times 2.64 = 1,596,672 \text{ ft}^3 \qquad [4.6]$$

If we made the pond 5 ft deep and we made it square, each side would be less than 565 ft. A football field is 300 ft between goal lines and 160 ft wide, so our pond is a bit larger than six football fields, not an impossible size. Our actual pond could be smaller because we probably would use less power after midnight, and because our stream would add water during the week. If we used power only 4 hours a day our pond could be about the size of a football field.

Actually what we mean by 5 ft deep is that the level of the pond would drop by 5 ft when water was drawn out. We would probably not want to empty the pond entirely, for the sake of the wildlife if nothing else, so the pond should be deeper than 5 ft.

The pond is used for storage. Another mode of storage is batteries. We could compare the cost of making the pond with that of buying the batteries, but we must factor in alternative use of the land and also the desirability of having a pond for recreation or similar purpose.

## *Dams*

The best advice about dams is that if you don't need to build one, then don't! One should try simply to divert water from a stream or try to take advantage of natural ponds or backwaters. If you are lucky enough to have an abandoned dam site on your property, try to reconstruct it rather than building on a new site. Regulatory agencies will be much more concerned about a hydropowered project if it contains a new dam.

In picking a site for a dam, one wants to maximize the head. One also wants to arrange the waterworks so that the generator, and thus the wheel, is close to the house to minimize electrical transmission losses. On the other hand, one probably does not want to live too close to either a wooden waterwheel or a turbine–generator combination, both of which can be noisy and vibrating. If one has a choice, one should avoid streams with silt, which can damage a turbine. Finally, as will be described next, leaks can be a problem so one needs to select a site for the pond and the dam that will have low leakage.

Some dams, which a person and friends could construct, are shown in Fig. 4.8. Dams can fail in two ways: (i) seepage underneath, which undermines the dam, and (ii) pressure that pushes the whole dam downstream. Seepage is minimized by using clay for the entire dam for a core or for a seal on the sides and bottom. The earthfill dam in Fig. 4.8 has a clay core. Pig manure is another sealant, as is plastic sheeting. Heavy dams are less likely to slide away than light ones, so rocks improve safety.

Figure 4.8 is probably self-explanatory. In an earthfill dam the slope of the upstream side should be about 1:3, so the water pushes down more than sideways, and large rocks should be used on the surface facing the water to avoid erosion. The slope of the downstream side can be steeper, perhaps 1:2. Rockfill dams can have a deeper slope than earthfill ones — perhaps 1:1.5. In the rockfill dam of Fig. 4.8, concrete is laid over the rocks to prevent leakage. The crib dam is made of logs stacked a few ft apart with gravel in between. Boards covering the face of the dam reduce leakage. Crib dams are best for low heights — 5 ft or less. The gravity dam is made of concrete and is the most expensive.

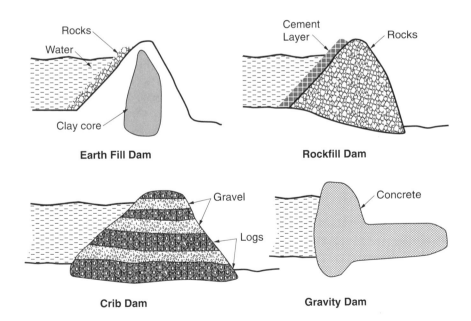

**Fig. 4.8. Types of dams.**

In any case, all dams need to have a spillway, where overflow water can safely get over the dam. It is probably clear that somebody will be in trouble if the dam washes away entirely, and floodwaters pouring over a dam can wash it away. It is also obvious that a way must exist to get the water from the pond to the head race, or the penstock. This water cannot be taken from the top of the pond because then the pond would only be useful when it was full. The water outlet should probably go through the dam. It will be easier to repair the dam if the pond can be completely emptied so a drain at the bottom of the dam is usually also included.

Two or three people can build small dams in several days. If necessary, one builds a temporary dam upstream to hold the water back during construction. The cost of construction depends on how much one does oneself. It can be small if one is industrious and has industrious friends. A reasonable estimate might be $150 for an earthfill dam built by the owner. On the other hand, a contractor might charge $2500 for the same dam.

## *Costs*

Step 3 in Fig. 4.3 concerns costs, which we discuss now. Three possible electrical systems are shown in Fig. 4.9.

Table 4.1 lists the costs of a typical, inexpensive system, (Fig. 4.9, top), which produces 1.500 kW. The 32-V electrical generator shown is designed to feed into the inverter, which permits the use of ordinary home appliances. The costs in Table 4.1 were estimated assuming the owner or her/his friends built the dam, foundation, and cross-flow turbine.

In this system the DC generator is the most vulnerable part and may need replacement, or at least an overhaul, in 5 years. A pond may silt up and should be cleared approximately every 10 years. Everything else should last about 20 years.

A simple estimate of annual costs, ignoring the cost of clearing out the silt, can be made by assuming the life of the system is 20 years. In 20 years, four DC generators will be needed. In our calculations we will assume all four are bought the first year, creating a capital cost of $4950. The capital recovery factor for 20 years of 12% is 0.132, so the annual cost for interest and principal is $653.

$$0.132 \times 4950 = \$653 \qquad [4.7]$$

The reader might compare the corresponding monthly cost, $54, to her/his own electric bill, which probably covers more than a 1.5-kW usage. Another comparison is the capital costs to the cost of bringing in electric power from the nearest utility line ($20,000 a mile). If the home is more than a short distance from the power lines, the hydropowered system is

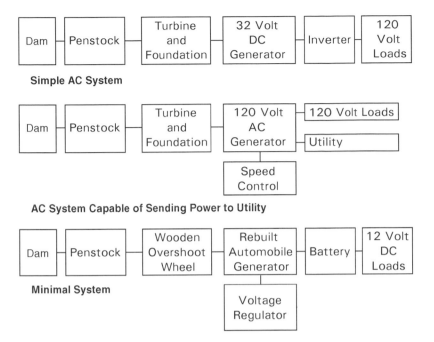

**Fig. 4.9. Electrical systems.**

cheaper, within a 20-year lifetime, especially when one considers the monthly electric utility bill in addition.

A more elaborate system (Fig. 4.9, middle) uses a high-quality AC 1.5-kW alternator producing 120 V directly. This alternator costs about $3000 and should last 20 years. If we were connected to the utility lines already and wanted to sell excess power back to the electric company, a speed control device is needed to maintain 60 cycles per second. This would cost an additional $1000.

**Table 4.1. Costs of a Small Hydropowered System**

| Component | Cost ($) |
|---|---|
| Dam | 150 |
| Penstock (200 ft) | 300 |
| Turbine and foundation | 100 |
| 32-V DC generator | 700 |
| Inverter | 1600 |
| Total | 2850 |

A minimal system (Fig. 4.9, bottom) might use a wooden overshot wheel with a wooden flume rather than a penstock. If scrap lumber is used, the costs would be much less than a turbine and PVC pipe. A bearing for the wheel would probably be necessary as well as some concrete for the foundation. These costs might be approximately $50. A rebuilt 80-A automobile alternator with a voltage regulator would give 960 W and cost about $40. It should last 5 years. The pond would supply most of the storage, but a deep-cycle battery would be useful to provide extra current when appliances are starting or when some other major current drain is occurring. A voltage regulator is also shown, an alternator will last longer with a voltage regulator. This minimal system would cost about $650. It, of course, would be more difficult to build and less reliable than the other two systems.

During certain years the U.S. government has given tax benefits to projects generating electricity. If these benefits exist, they also should be factored into the costs calculations. The last two steps in the design process of Fig. 4.3 are financial and legal.

## Legal Requirements

Several agencies have jurisdiction over hydroelectric installations and the laws differ state by state and possibly county by county. Therefore, it is advised that one must be careful. Noncompliance will probably mean removal of the installation and possibly a fine. The state agency dealing with the environment will probably be interested in your project. So will the fisheries department. The Federal Energy Regulatory Commission licenses hydroelectric projects. Normally obtaining such a license is no job for an amateur; however, an exemption is usually granted if the project is <100 kW the typical case for a small-scale project. The local building inspector should pass on construction plans.

A lawyer experienced in such matters should probably be consulted early. It often takes a year to get permits. One also needs to check who actually owns the water rights to a stream. They do not always go with the property. The concerns of neighbors downstream should be taken into account. After a dam is built, at the very least, they may be getting their water at different times of the day, or week, as you open or shut the sluice gates.

Financing can be arranged through a bank, of course. Regional offices of the Department of Energy may also be helpful because the federal government has programs supporting small-scale hydroelectric projects. The Report from the National Center for Appropriate Technology [5] describes both financial and legal matters. Before one actually builds anything or spends any money one should definitely check with a lawyer.

## ENVIRONMENTAL CONSIDERATIONS

Ponds change streams and thus change the ecology. They provide habitats for some varieties of fish and reptiles and store water in the dry seasons for mammals such as deer or otters. On the other hand, they destroy areas of running water which trout and salmon need. Dams are a major obstacle to fish returning from the ocean to spawning grounds at the headwaters of streams. State government agencies concerned with fishing will nearly always insist on "fish ladders," which are basically small pools arranged in steps so fish can jump from one pool to another. Ladders, however, may not be enough because the pond behind a dam may not be a proper spawning place for a fish used to spawning in running water.

The wheel or turbine itself can do environmental harm. Live fish, and other creatures, can be hurt in the blades of a turbine or, less likely, in the buckets of a water wheel. They can also cause damage to the turbine or waterwheel, so the trash racks used to keep sticks out of the penstock or head race should be fine enough to keep fish out as well.

Other environmental effects need to be considered. A major, and obvious, concern is the integrity of the dam. Noise and the visual impact can be important. A wheel creaks and a generator vibrates. The noises are not objectionable if one is 30 ft or farther away, but if living quarters are closer, soundproofing may be needed. One has to use one's judgment about the visual impact. Waterwheels have had a long history in the United States and are often an attraction for visitors.

This point, the historical associations of water power, is worth amplifying. Many cities and towns in the United States developed at their current locations because water power was available. Colonial farmers built dams and waterwheels chiefly to grind corn and other grains. Water power is an integral part of the heritage of the United States, beginning with individual wooden wheels on farms and continuing during the textile mills built in the nineteenth century. Reconstructing or replicating this water-driven machinery preserves an important part of U.S. culture.

## HYDRAULIC RAM

Another way to use water power deserves mention, although it cannot be directly used to generate electricity. A hydraulic ram, shown in Fig. 4.10, is a simple device which uses the energy in a stream with high volume, but

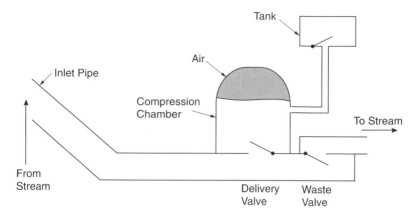

**Fig. 4.10. Hydraulic ram.**

low head, to lift a small volume of water to a large height. It could be used, for example, when we have a stream with a 1-ft drop carrying 100 cubic ft per minute of water and want to raise 2 cubic ft of water 8 ft to a storage tank on a roof to supply a kitchen and bathroom with running water.

The only part of the ram in Fig. 4.10 which is not obvious is the two valves. They are made with weights which normally open or close them, depending on the position of the weight. In particular, the delivery valve in Fig. 4.10 is normally closed, i.e., horizontal, but can open when water pushes from below. It closes when water pushes from above. Similarly, the waste valve in Fig. 4.10 is normally open, i.e., vertical, but closes when water pushes up from the bottom.

It is probably easiest to understand how the hydraulic ram works by describing the operation step by step. Initially, the delivery valve is closed and the waste valve is open:

1. Water from the stream flows down the input pipe, forcing the delivery valve open and the waste valve closed. Thus, water can only flow into the compression chamber.

2. The momentum of the large amount of water from the input brings much water into the compression chamber, compressing the air on top.

3. Because the water is flowing into the chamber, the pressure on the waste valve is small, so the waste valve falls open.

4. When the air in the compression chamber is sufficiently compressed it pushes back on the water, closing the delivery valve.

5. The compressed air pushes on the water in the chamber, forcing it up to the tank.

6. Simultaneously, the water from the input, having no other place to go, flows out the waste valve, pushing it closed.

7. Meanwhile, the pressure on top of the compression chamber has decreased, as water goes to the tank, so the delivery valve opens.

The process repeats itself, with the situation being the same as in step 1: delivery valve open and waste valve closed.

Hydraulic rams essentially have no parts that rub against each other, as many pumps have, so they last a long time — several decades at least.

## REFERENCES

1. Center for Alternative Technology (1998). D.I.Y. Plan 7 — Timber Waterwheel (pp. 7). Mackynlletts, Powys, Wales.

   *Working plans with construction notes, seems complete and doable.*

2. Leckie, J., Masters, G., Whitehouse, H., and Young, L. (1981). *More Other Homes and Garbage* (pp. 75–92). Sierra Club Books, San Francisco.

   *Especially thorough on system design and measuring stream flow. Design formulae for an overshot wheel. The hydraulic ram is discussed on pages 313–314. Some supplier and cost information is given.*

3. Mother Earth News (1983). *Mother's Energy Efficiency Book* (pp. 86–105). Mother Earth News, 105 Stony Mountain Road, Hendersonville, NC 28791.

   *It contains working drawings for a hydraulic ram and a small working overshot wheel. An informal discussion of much of what is in* More Other Homes and Garbage.

4. Mother Earth News (1981), Mother's Crossflow Turbine Plans. See address in Ref. [3] ($15).

   *Working drawings seem possible for the amateur but certainly not a trivial project. The manual is very helpful; mostly system design — measuring flow, determining dimensions, and electric components needed. List of suppliers.*

5. National Center for Appropriate Technology (1979). Micro-Hydropowered Power (DOE/ET/01752-1). U.S. Department of Energy, Washington, DC.

   *Very thorough guide to determining feasibility and then designing a system. Discusses regulatory process. Many tips on how to do things. Somewhat out of date, but best place to start if you really want to build a hydropowered installation.*

6. Real Goods (1991). *Alternative Energy Sourcebook Real Goods.* Real Goods, 966 Mazzoni Street, Ukiah, CA 95482. Telephone: 707/468-9214 ($10).

   *Lists turbines and generators, as well as several pages of design information.*

# PROBLEMS AND PROJECTS

1. This problem concerns the amount of power available from a stream.

    A. If one wanted 2000 W and a head of 12 ft were available, about how much water in cubic ft/second would be needed? Assume an efficiency of 50%.

    B. If one wanted 5 kW and the penstock could deliver 7 ft$^3$ of water/second, how much head is required? Assume an efficiency of 70%.

2. What width and height for a flume are necessary to deliver 3 ft$^3$/s of water if the velocity of the water is 2 ft/s?

3. (A) Estimate the torque produced by an overshot waterwheel. Assume each bucket holds 10 lbs of water, when holding water, and that seven buckets are holding water at any given time. The diameter of the waterwheel is 16 ft, measured to the middle of the buckets. Torque is measured in foot-pounds. (B) A 0.5-horsepower motor, as used in small grinding machines, drill presses, lathes, and so forth, produces about 9 ft-lbs of torque. How many such motors could our waterwheel replace, even if half the torque is wasted in the gears and belts connecting the wheel to the machine?

4. Consider a stream or river that you know about or hypothesize one. (Many parks in urban areas have streams, if you have difficulty hypothesizing.) Lay out a hydro-electric installation, showing where the races or the penstock will be. Include approximate dimensions. Estimate the power obtained.

5. How big a pond is needed to supply 2 kW for 100 hr if the head is 10 ft? Assume the highest dam feasible is 8 ft high and that the efficiency is 10%.

6. Assume you have a place on your property where a stream passes through a gully which is 4 ft wide at the bottom and 8 ft wide at the top. The gully is 10 ft deep. Estimate how long it would take you and a friend to build an earthfill dam in the gully. Start by estimating how much earth is needed. Assume clay is available at the site.

7. (A) Cost out the more elaborate AC generating system of Fig. 4.9, producing 1.5 kW. (B) Assume you built the system of part a) and only use 6.000 kW-hr a day (1 kW for 6 hr a day). The stream powers the generator 24 hr a day. You will sell your excess power to the electric company at $.08 per kilowatt-hour. How much money will you make a year?

8. People who enjoy white-water canoeing do not want to see fast-moving rivers dammed. Assume you were one of the canoeing enthusiasts and were to make a presentation at a Department of Environmental Management hearing on a proposal to build a small-scale hydropowered electric plant. What arguments would you use?

9. Consider the same situation as in Problem 8, but assume your major concern was encouraging use of renewable energy sources. What arguments would you use?

*The next three projects involve actual construction.*

10. Build a working model of an overshoot waterwheel.

11. Build an actual cross-flow turbine from commercial plans (see References) and test it.

12. Build a small hydraulic ram and test it.

*Chapter*

# 5

· · · · · · · · · ·

# WIND POWER

· · · · · · · · · · · · · · · · · · · · ·

## WHY WIND POWER?

Wind power has a large advantage over water power in that although not everyone has a stream, nearly everyone has access to wind. Before the Rural Electrification Act in the early 1930s, nearly all farms in the Midwest got their power from the wind. Cheap electricity has made wind power less desirable but when oil prices rise significantly again wind energy will make more sense, not only in rural areas but also for individual suburban homes. Although it clearly would not be feasible for everyone in large apartment building to set up individual windmills, a few city homeowners have mounted windmills on their roofs and sold electricity back to the utility. Wind power is also still used in rural areas. Commercial wind machines, popular before rural electrification, are still sold in the United States; a list is given in Refs. [2] and [8].

Of course, the wind does not blow all the time, so storage is necessary. If one is using the wind to generate electricity, then batteries are usually the storage means, unless one can use the utility company for storage, as described in the chapters on electricity and water power. On the other hand, some important applications, particularly irrigation or refrigeration, do not require storage. Probably the primary use of wind power on farms has been pumping water for livestock or crops, an application that matches the capabilities of wind machines well. Because electricity seems to be a more general need, we focus on it in this chapter, even though it is probably not the optimum application for wind.

## THE PHYSICS OF WIND POWER

Wind power originates in the kinetic energy of the wind, that is, the energy the wind has because of its motion. Nearly all wind-power generators use a rotating propeller that captures the kinetic energy when the wind pushes on the propeller blade. The reader may recall that kinetic energy of an object equals one-half its mass times the square of its velocity. In this case the object of interest is a block of air which goes through the blades. Power is energy per second, and the amount of the air reaching the blades per second depends on the velocity of the air, so the power depends on the velocity cubed. The power also depends on the size of the block of air captured by the propeller, that is, on the area covered by the propeller:

$$P = V^3 \times A \times e \times 5.3 \times 10^{-6} \text{ kW/mph ft}^2 \qquad [5.1]$$

Where $P$ is the power out in kilowatts, $V$ is the wind velocity in miles per hour (mph), $A$ is the area swept out by blades in ft$^2$, and $e$ is the efficiency of propeller. The factor $5.3 \times 10^{-6}$ arises because of the units chosen. Actually, this factor depends somewhat on the density of the air—the number of pounds a cubic foot of air weighs—as the mass of a cubic foot of air depends on its density. We will, however, ignore the small change in density caused by changes in temperature and altitude—hot air is less dense than cold air, as is air at high altitude. Efficiencies for various kinds of propellers are given when propellers are discussed.

One implication of Eq. [5.1] is that power increases rapidly as the wind speed increases. When the wind speed doubles the power goes up eight times, or a 25% increase in wind speed doubles the output power. This variation of power with wind speed is important because wind often blows in gusts, as shown in Fig. 5.1.

Most of the power occurs at high velocities rather than at the average velocity. In other words, use of the average wind speed in Eq. [5.1]. will give a wrong estimate of the power. Two winds may have the same average speed but one may have gusts at higher speed and thus higher power (see Problem 3). It is much easier to measure the average wind speed than the variations in speed, and the National Weather Service reports average wind speeds. A correction factor is needed in Eq. [5.1] to get an accurate estimate of the power when the value for the wind velocity, $V$, is the average wind speed. This correction factor equals 2.5; that is, the power calculated from Eq. [5.1] needs to be multiplied by 2.5 if the speed, $V$, is the average wind speed.

Wind speed increases with height, which makes sense because the wind speed at ground level is 0. The increase is described approximately by Eq. [5.2], which is based on measurements. (Actually, the proper exponent in Eq. [5.2] depends on the terrain, but we ignore this dependence.)

$$\frac{V_1}{V_2} = \left(\frac{h_1}{h_2}\right)^{0.2} \qquad\qquad [5.2]$$

By substituting values into Eq. [5.2], we can see that increasing the height of a tower from 5 to 30 ft will increase the velocity by a factor of 1.43. This would approximately triple the power output.

**Fig. 5.1. Variation in wind speed.**

Estimating the wind speed is an important matter in designing a wind generator. The National Weather Service gives average wind velocity for each of their stations — usually at airports. *The Wind Power Book*, [7] gives these averages by month. Wind, though, is very local — sites even 100 ft apart may have significantly different winds, so one has to be careful in using their values. One approach is to compare a specific site with that of the Weather Service and estimate the similarity in wind conditions. Another approach is simply to measure the wind. Commercial equipment is available or the *More Other Homes and Garbage book* [3] describes how to make such measuring equipment. Of course, careful measurements, even over a long time, will give only estimates and not perfect forecasts because the wind is not reliable. The descriptions in Table 5.1 can be helpful but are not really accurate.

## Table 5.1. Beaufort Scale of Wind Force

| Condition | Observations | Speed (mph) |
|---|---|---|
| Calm | Calm; smoke rises vertically | 0–1 |
| Light air | Direction of wind shown by smoke drift but not by wind vanes | 1–3 |
| Light breeze | Wind felt on face; leaves rustle; ordinary vane moved on wind | 4–7 |
| Gentle breeze | Leaves and small twigs in constant motion; wind extends light flags | 8–12 |
| Moderate breeze | Raises dust and loose paper; small branches are moved | 13–18 |
| Fresh breeze | Small trees in leaf begin to sway; crested wavelets form on inland waters | 18–24 |
| Strong breeze | Large branches in motion; whistling in telegraph wires; umbrellas used with difficulty | 25–31 |
| Near gale | Whole trees in motion; inconvenience is felt when walking against the wind | 32–38 |
| Gale | Breaks twigs off trees; generally impedes progress | 39–46 |
| Strong gale | Slight structural damage occurs (chimneys and roofs) | 47–54 |
| Storm | Seldom experienced inland; trees uprooted; considerable structural damage occurs | 59–63 |

# TYPES OF WIND MACHINES

Two types of wind machines exist: the "drag" and the "lift" types. A typical example of each is shown in Fig. 5.2.

The drag-type rotates because the wind pushes against it. Square-rigged sailing ships move because of drag. The rotor in Fig. 5.2 is propelled because the wind is caught by one surface and flows over the other. It is called Savonius after its inventor. It rotates about a vertical axis, and it has the advantage of high torque, especially when starting because the wind has a large area to push against. It has a disadvantage at high speeds because the wind pattern is badly disrupted (turbulence is produced), so the efficiency declines. Also, the curved surfaces cannot move faster than the wind because they are being pushed. The details of a Savonius rotor will be discussed later.

The lift type, the other rotor of Fig. 5.2, is more efficient at high speeds because it produces less turbulence, but it has a low torque and thus can be hard to start under load. The principle of operation, which is similar to that of a sailboat or airplane wing, is illustrated in Fig. 5.3.

When the wind blows by a blade, the wind "tries" to keep going in the same direction. The result is that on the bottom of the blade, the wind is concentrated near the blade, whereas on the top it is away from the blade. Therefore, the region below the blade is at high pressure and the region above the blade at low pressure, so the blade is pulled, or lifted, as shown. Sailboats, unless they are going downwind, move by the same principle: The wind is deflected to produce low pressure on one side of the sail and high pressure on the other side. The pressure difference produces a force.

An essential consideration in dealing with wind machines is what happens when the wind speed is very high. A way of preventing the rotor from going too fast is needed. A mechanical brake can be used but is cumbersome. The blade angle of a lift-type device can be adjusted to eliminate the lift when overspeeding occurs. A horizontal axis machine can be rotated so it does not face the wind. Usually this rotation is sideways so the wind blows on the end of the blade rather than the front, but it can be vertical, tipping the rotor so it faces up. This concern for high wind speeds is an example of the importance of considering not only expected conditions but also extremes in the designing process.

## Savonius Rotor

Top views of two Savonius rotors are shown in Fig. 5.4. The model shown in Fig. 5.4a is simple to make. It can be constructed by slicing an oil drum

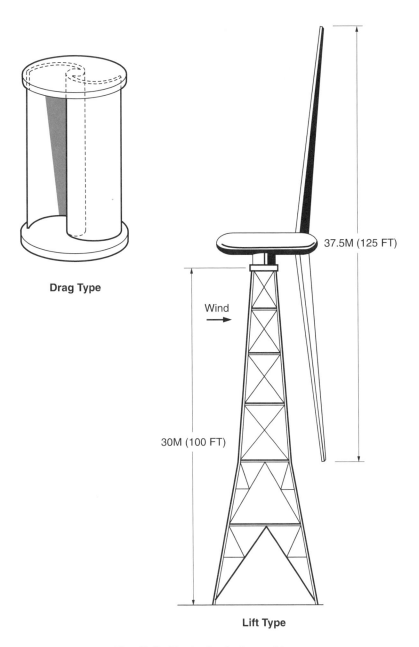

**Drag Type**

Wind

37.5M (125 FT)

30M (100 FT)

**Lift Type**

**Fig. 5.2. Typical wind machines.**

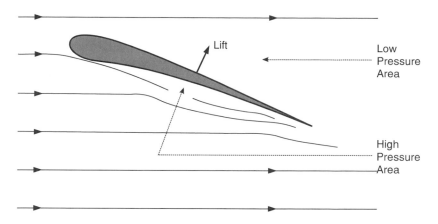

**Fig. 5.3. How lift is created.**

down the central axis and welding the two halves to form an S-shaped rotor. The model in Fig. 5.4b is somewhat more efficient because the shape gives some lift when the blade faces the wind. Most of the torque produced, however, comes from the wind pushing on the concave surface. Of course, the wind exerts the most force when it flows directly into one of the pieces, so the torque produced varies as the rotor spins. This variation in torque can be reduced if three or four Savonius-type rotors are stacked on top of each other, offset in angular position.

Savonius rotors are easy to build. The student project shown in Fig. 1.7 is a Savonius rotor. As indicated previously, they produce high torque even when starting, so they are useful for grinding and other machinery. They are often used for pumping water. On the other hand, they are not

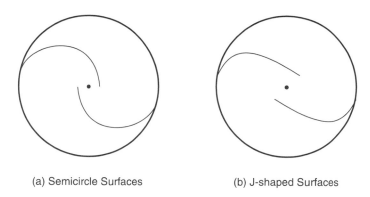

(a) Semicircle Surfaces          (b) J-shaped Surfaces

**Fig. 5.4. Savonius rotor.**

efficient — about 15% when coupled to a load. It is not necessary to face them into the wind because they always do. If the wind direction is changing rapidly, not having to move the rotor to face the wind is a big advantage. The vertical axis may mean that gears are not needed to bring the power down from the tower top. One does not have to worry about overspeed conditions because as high winds push against the other surface and keep the speed under control. A disadvantage of Savonius rotors for electrical power generation, though, is that the low speed probably means that gears or pulleys must be used to match the rotor to a generator because low-speed generators are not easily available.

## Propeller-Type Rotors

Propeller-type rotors (Fig. 5.2), are high revolutions per minute (rpm) devices, so they are ideal for electric generators. The efficiency is high — 40% for the entire system at high speed, but lower at lower wind speed. One problem is that the optimum angle of attack of the blade depends on the wind speed. Propellers are usually designed for high speeds, which reduces the efficiency at low speed, although some have intricate mechanisms to control the angle of attack or pitch.

Some design problems do exist. If the rotor is upwind from the tower, then some mechanism is needed to aim the rotor into the wind. Usually a tail is used. An advantage of a tail is that a mechanism can be included which will face the rotor out of the wind when the velocity is too high. If the rotor is downwind from the tower, as in Fig. 5.2, a tail is not needed, but the rotor blades get less wind when behind the tower (in the shadow of the tower). A downwind rotor will probably vibrate as the blade goes behind the tower.

The exact shape of the blades is important for high efficiency. If you are building it yourself, you should get a good set of drawings, for example, the Mother Earth plans [4]. The design problem for a propeller blade is that the actual linear speed, when the blade is turning, increases along the blade. The end of the blade is moving much faster through the air than the parts near the center, so the propeller must be cut with a varying pitch. Some moderately skilled woodworkers have built successful propellers, but the task is not simple, especially if a large blade is needed.

## Sailwing Rotors

A variation on the propeller rotor is the sail wing (Fig. 5.5). The sail wing can be built cheaply by an amateur. The blades are made of cloth sails.

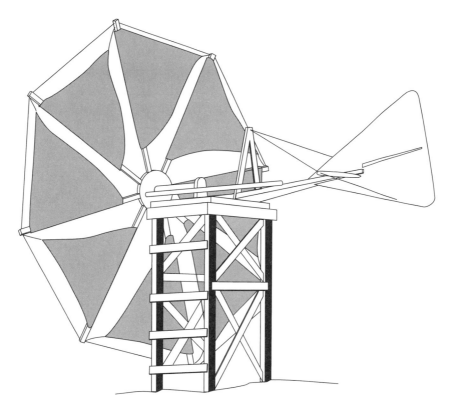

**Fig. 5.5. Sail wing rotor. (Reproduced with permission from National Centre for Appropriate Technology [5].)**

They can be connected to the hub with springs so that in high winds they automatically feather (face the wind directly, thus losing lift.) The New Alchemy plans [6] give details. Either cotton or Dacron can be used for the sails and should last 3 or 4 years (compared with 80–100 years for a wooden propeller). Wide sails give high torque, and narrow sails give high speed. Sail wings, therefore, can be used for both pumping and electricity, but not for both at once. They have about 15% efficiency.

Even with springs to feather the blades at high wind velocities, one has to be careful when storms occur. At velocities over 30 mph it is necessary to furl the sails to avoid damage. It is not prudent just to hope that the sail wing will get through a storm without adjustment.

## SITING

If a wind machine is used for water pumping or for other mechanical devices then the machine needs to be close to where the power is used. If the machine is to generate electricity, one has a greater choice of sites. As shown in Eq. [5.2], the higher the location the better. In any case, the rotor should be at least 10 ft above wind obstacles within 100 yards.

Tops of gently sloped hills, as shown on the left in Fig. 5.6, are quite suitable. The wind climbs the hill smoothly and is concentrated at the top. Steep hills, as shown on the right in Fig. 5.6, are less suitable since turbulence occurs at the corners. Tops of buildings tend to be undesirable for the same reason in addition to the difficulty of building a stable mounting for the tower. A valley aligned parallel to the direction of the wind can channel and concentrate it. A commercial experiment in Altamont, California, near San Francisco, uses many wind generators located in such a valley to produce electricity for a utility.

## TOWER AND TRANSMISSION

Towers can be homemade, but unless one has some structural engineering skills it may be best to purchase one. The cost is between $500 and $1000. Generators producing the most energy, as you might expect, need the strongest towers because they slow down the most wind. Towers for high-speed commercial generators, therefore, should be purchased. Towers are especially stressed when the wind changes either speed or direction because of the twisting, so at a site with erratic winds a particularly strong tower must be used. Towers deserve serious thought because the consequences of a falling wind machine can be disastrous.

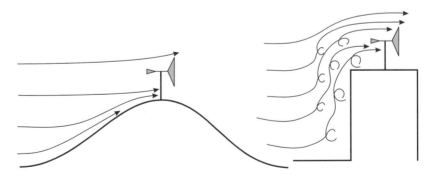

**Fig. 5.6. Suitable and unsuitable wind machine sites.**

"Transmission" refers to the means for getting power down from the rotor to the ground. If the wind machine powers an electric generator one solution is to put the generator up on top of the tower connected directly or through gears to the rotor, as is done in many commercial wind generators. In this case the tower must support the additional weight and servicing the generator on top of a tower can be difficult. If the generator is on top, electric wires must, of course, be brought down to the ground, so it is necessary to prevent these wires from becoming tangled and breaking. An elegant solution is to use slip rings and brushes. Slip rings are two rings of brass, each encircling the turntable on which the generator is mounted. The generator output is connected to the rings. Stationary carbon bars, called brushes, rub against the rings. No matter which direction the rotor is facing, current can flow through the rings to the brushes and then down through wires to the ground. If one is willing to keep one's eye on the wind machine, however, such a solution may not be necessary.

If the generator is on the ground or if the windmill is being used to drive a mechanical device, then a mechanical transmission is needed. Often these transmissions serve the additional purpose of increasing the speed of rotation above that produced by the rotor. It is useful to be able to increase the speed of drag-type machines, which tend to rotate much slower (100 rpm) than the rated speed of a generator (3600 rpm). If the wind machine has a horizontal axis, like a propeller or sail wing, then the transmission must also change the direction of the shaft from horizontal to vertical.

An automobile differential in a rear-wheel-drive car does just that. It uses the driveshaft, which extends back from the engine, to drive the rear axle. (The axle is at a right angle to the drive shaft.) The automobile differential can also handle the motion of the rotor housing as the wind direction changes. If the differential is properly installed (actually opposite to the way it is set in an automobile, where the engine goes faster than the wheel), it will make the shaft going down the tower rotate faster than the rotor.

## TRANSMISSION SYSTEM DESIGN

To see how much the rotor speed must be increased by the transmission system, we first estimate how fast the rotor will turn. We use the tip:speed ratio, which is the ratio between the speed of the tip of the blade and the wind speed. This has been worked out theoretically and measured experimentally. For a sail wing it is about 5, for a propeller as high as 8, and for a Savonius rotor about 1.

Assume we have a sail wing and want to know how fast the rotor revolves. If the wind velocity is 10 mph and the tip:speed ratio is 5, then the tip is moving at 50 mph or 4400 ft per minute. If the length of a sail is 10 ft, then a revolution is $2\pi \times 10$ ft, so the number of revolutions per minute is

$$\text{Revolutions/minute} = \frac{4400}{20\pi} = 70.1 \text{ rpm} \qquad [5.3]$$

If we want to drive an automobile generator at about 1000 rpm, we need to gear up about 14 times. Automobile gear boxes can be modified to give this increase in speed but the modification is somewhat difficult. Commercial gearboxes, which can also be purchased, are lighter and probably more reliable than scrap automobile parts. Tractor stores sell gearboxes for about $100.

Rather than a vertical shaft and gearbox, one could use a chain drive (like a bicycle) or even a belt drive. Chain drives are heavy and require regular oiling and tightening. Belts and pulleys are the cheapest solution but waste some power because they slip. Timing belts have grooves and do not slip but cost more and need matching pulleys.

## ELECTRIC GENERATION AND STORAGE

Three wind-driven electric generating systems are shown in Fig. 5.7. In Chapter 3, we considered storage of electricity, which is best done in batteries in a simple system and either in batteries or with the utility in a larger system. If we are to use batteries we need direct current and,

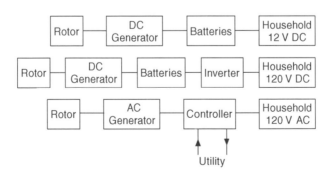

**Fig. 5.7. Some alternative wind electric systems.**

because the batteries are 12 V, we should generate at 12 V. Actually, higher voltages are possible because batteries can be connected in series, so we could store 24 V, 36 V or any multiple of 12 V. Appliances normally run on 120 V AC; however, some appliances are available for 12 V DC, so a 12-V system makes sense. The top system in Fig. 5.7 shows a 12-V generator charging batteries which feed 12 V DC appliances. Prices of 12-V DC appliances were given in Chapter 3.

If we prefer to use 120 V AC appliances, which are more common and thus easier to obtain, we could use an inverter to go from 12 V DC to 120 V AC, as in the middle system in Fig. 5.7. An advantage of 120 V over 12 V is that for the same power a 120 V system uses one-tenth the current, so losses in the connecting wires are less. Another way to get 120 V AC is to buy an AC generator. Many AC generators have speed controls so the voltage produced is exactly 60 cycles per second. The output from these generators can be interchanged with the utility (Fig. 5.7, bottom). The utility here is always connected to the controller. When the wind is blowing strongly the AC generator produces excess power and sends some to the utility and the rest to the household loads. When the wind is not blowing strongly the household gets its electricity from the utility. In this system, the wind generator's function is to reduce the household electric bill.

## DESIGN AND COSTS

We will make a design and estimate costs for the simplest system in Fig. 5.7 and compare those costs with the purchased system. Cost, of course, depends on how much one does oneself. The top system in Fig. 5.7 can be built almost entirely by hand so costs are basically for components and supplies. We will use a sail wing rotor. The DC generator could be an automobile alternator with rectifiers and a voltage regulator. We assume that we need enough storage for 4 days without wind; that is, we want to be able to charge our batteries in 24 hours with enough energy to supply the family in case wind does not blow for the next 3 days. We will consider the same family of four people as in Chapter 3 who need 960 W for 5 hours each day; this family uses electricity only in the evenings.

First, we calculate how big a rotor is needed to charge the batteries during 1 day to supply our home for the next 4 days. In other words, during 24 hours we want to recharge the batteries so they will give 960 W for the next 5 hours and for 5 hours during each of the next 3 days.

We use Eq. [5.1]. We assume the average wind velocity reported by the Weather Service is 10 mph, and we use a factor of 2.5 to account for the ratio between the average power and the power calculated from the average velocity:

$$P_{needed} = V^3 \times A \times e \times 5.3 \times 10^{-6} \qquad [5.4]$$

where $P_{needed}$ is the power in kilowatts needed from the generator during the 24 hours the wind is blowing, $V$ is the wind velocity (10 mph), $A$ is the area swept by rotor, and $e$ is the efficiency for a sail wing rotor (15%).

The energy needed to get 960 W for 5 hours a day for 4 days is

$$\text{Energy needed} = 960 \times 5 \times 4 = 19{,}200 \text{ W-hr} \qquad [5.5]$$

The generator must supply this amount of energy in 24 hours. The power, $P_{needed}$, required from the generator is

$$P_{needed} \times 24 = 19{,}200 \text{ W-hr}$$
$$P_{needed} = 800 \text{ W} = 0.8 \text{ kW} \qquad [5.6]$$

Now that we know $P_{needed}$ we can solve Eq. [5.4] for the area swept out by the blades of the rotor. We find the area to be 402.5 ft$^2$. The corresponding blade length is 11.3 ft $- A = \pi r^2$, where $r$ is the blade length.

An 11-ft blade is somewhat long, so we might decide to use two rotors, towers, and generators. In this case the blades of each would be 8 ft long. To avoid power loss, wind machines should be separated from each other by at least a distance of 15 times the rotor diameter, or about 120 ft in our case.

We should also check how much current the generator must supply to the batteries when they are recharging. Remember, power equals voltage times current:

$$P_{needed} = 12 \times I$$
$$I = 66.67 \text{ A} \qquad [5.7]$$

An automobile generator this large is uncommon but we could use a small truck alternator. If we cannot find a big enough alternator, we could drive two smaller alternators with the same rotor.

It was shown in Eq. [5.3] that the rotor revolves at about 70 rpm. An automobile generator needs to be going at least 500 rpm to produce rated current; 1000 revolutions would be better. Therefore, we need to increase the speed by a factor of at least 8. As previously discussed, pulleys and a belt are the cheapest method, although gears are more reliable. We decide to mount the rotor on top of the tower and use slip rings brushes to connect its output to the supply wires.

Blades 11 ft long would require a tower at least 21 ft tall, and one even taller would be better. We choose 30 ft — approximately the height of a three story building. We certainly would not want to build higher without professional help. The design of the rotor and tower is now complete.

Now we estimate the number of batteries needed. We use deep-cycle batteries costing $80 each and draw 9 A from each. According to the manufacturer, the batteries in this case will last 10 hours before needing a recharge. The energy stored in each 12 V battery is the power — volts × amperes — times the hours used.

$$\text{Energy per battery} = 12\text{ V} \times 9\text{ A} \times 10\text{ hr} = 1080\text{ W-hr} \qquad [5.8]$$

The energy needed for 960 W for 5 hours a day for 4 days is (using Eq. [5.4]) 19,200 W-hr. The number of batteries needed is 18.

Now that we know the size of all the components, we can estimate costs. These are given in Table 5.2, in which we have decided to use only one tower. The biggest component of the cost is the batteries. If we are in a location where the wind is dependable and we only needed 1 day storage, five batteries would be enough. In this case the total cost would be $840.

By comparison, a so-called "plug-in" AC system, one that is ready to operate once erected, would cost $3300 plus $1800 for the tower, or a total of $5100 [2]. Such a system uses the utility for storage, so it only makes sense when the power lines are already connected.

### Table 5.2. Cost of Simple Wind System

| Component | Cost ($) |
| --- | --- |
| Rotor: sailing, 11-ft long sails | 250 |
| Generator: automobile alternator, 67 A | 50 |
| Tower: Wooden, handmade, 30 ft lumber | 100 |
| Batteries: Deep-cycle type | 1440 |
| Pulleys and belt: 8 : 1 ratio | 20 |
| Supply wire: Generator to ground 50 ft | 20 |
| Total | 1880 |

## Environmental Concerns

The major environmental degradation comes from noise. Windmills tend to hum or whine, and several together can be annoying. Another annoyance is that metal blades interfere with television reception. The television signals bounce off the blades and into the TV intermittently, so a disturbing pattern is superimposed on the picture. On Block Island, off Rhode Island, a big wind generator was installed to serve the local utility. To prevent TV problems the whole island was connected to cable TV. Another approach to eliminating interference is to use wood, plastic, or cloth blades, which do not cause TV interference but do not last as long as metal ones.

## Usefulness of Wind Power

Wind power certainly makes sense if one is in an isolated spot far from power lines. (In this context "far" means 0.5 mile or so because the cost of bringing in power lines is $20,000 a mile.) Of course, the site must have unobstructed access to the wind and the wind must be dependable and at least 10 mph on average.

On a site where power from the utility is already available, utility electric prices would have to approximately triple before power from a wind generator would be cheaper. Electricity prices depend on the price of oil which has varied greatly in the past few years, so tripling is not out of the question. Use of a wind generator saves coal and oil so it benefits society as well as the owner. While "visual pollution" is a matter of taste, some people find a wind generator attractive.

## References

1. Eldridge, F. (1995). *Wind Machines*, National Science Report (Stock No. 038-000-00272-4, pp. 77). Available from Superintendent of Documents, U.S. Government Printing Office, Washington, DC 20402.

   *Really focuses on big commercial systems; has some design data; mentions briefly many ideas not appearing elsewhere; good for policy considerations.*

2. Gavin, B. N., Proprietor, *Wind Electric Shop Catalogue and Price Lists*, Maple Avenue, Little Compton, RI, 02837.

   *Very helpful source of information. The company carries the well-known Jacobs line — probably the best known commercial wind generator.*

3. Leckie, J., Masters, G., Whitehouse, H., and Young, L. (1981). *More Other Homes and Garbage*. Sierra Club Books, San Francisco.

*The wind section is 26 pages with much design data. A device for measuring wind speed is described. There probably is not, however, enough information here to design and build a system yourself from scratch.*

4. Mother's Plans, P.O. Box 70, Hendersonville, NC 28793.

   *Has detailed plans for a wind plant, including drawings of a propeller ($15 plus $1 shipping and handling). A student made a propeller with good results from an earlier version of these plans.*

5. National Center for Appropriate Technology (1977). D.I.Y. Plan 5 — Sail Windmill Machynlleth, Powgs, Wales.

   *Has several pages on wind power and plans for a Cretan sail wing.*

6. New Alchemy Institute, *The Journal of the New Alchemists*, Woods Hole, Falmouth, MA 02543.

   *The New Alchemy Institute has done much with homemade windmills and commercial units. Plans and detailed constructions for a sail wing are given in Technical Bulletin No. 1. and in Vols. 2 and 3 (1974 and 1976). Experiences building a Savonius rotor are given in Vol. 3. Volume 2 reviews the electrical system as relevant to windmills. An article in Vol. 6 is a good place to start if you are interested in pumping water. Finally, Vol. 3 has some pictures of approximately 50-year-old commercial units — especially interesting to industrial archeology buffs.*

7. Park, J. (1981). *The Wind Power Book*. Chishere Books, Palo Alto, CA.

   *Probably the most complete and helpful reference. Discusses construction anecdotes, many design graphs, and Weather Service data. Definitely the place to start if you want to build a system.*

8. Real Goods, *Alternative Energy Sourcebook 1991*. Real Goods, 966 Mazzoni Street, Ukiah, CA 95482.

   *Four pages on wind systems. A 12-V 250-W generator is listed selling for $795.*

9. Rousmaniere, J. (1987, February/March). Soaring on the sea. *Air and Space*, **1** (6).

   *Explains how sailboats use the "lift" principle. Gives the aerodynamics helpful for building windmill rotors.*

## PROBLEMS AND PROJECTS

1. Estimate the power available from a propeller with a diameter of 10 ft when the wind speed is 10 mph. Assume an efficiency of 40%.

2. Estimate the power available from a Savonius rotor 4 ft high and 3 ft in diameter when the wind speed is 12 mph. Assume an efficiency of 15%.

3. The point of this problem is to determine the ratio between the average of a number cubed and the cube of the average. Assume the wind velocity is 17 mph half of the time and 8 mph the other half the time, so the average wind speed is 12.5 mph. Compute the average of $17^3$ and $8^3$ and compare it to $12.5^3$.

4. Draw some sketches and show how a sailboat can sail almost into the wind.

5. Figure 5.8 shows a top view of a sailboat with a jib. Explain why the jib makes the boat go faster. (Hint: When air is moving rapidly in a horizontal channel the vertical pressure on the channel walls is reduced compared to when the air is still.)

6. Would you consider a house fan a lift or drag device? How about a paddle wheel on a boat?

7. Invent shapes for rotors, besides the Savonius, that are drag devices.

8. Sketch a multitier Savonius and explain why it gives a more constant torque than a simple rotor.

9. Invent a lift device with a vertical axis. (Actually, this has been done and the device is called a Darrieus rotor.)

10. How might you prevent a downwind rotor from overspeeding at high wind velocity?

11. Consider the locality where you live and select the site which probably would have the best wind conditions.

12. Estimate the number of revolutions per minute from a Savonius rotor in a 15 mph wind. The rotor radius is 3 ft.

13. A friend asks you, "Why not generate at 120 volts AC to begin with, rather than use an inverter?" Why not, indeed?

14. How many batteries should be purchased if we desired to allow for 6 days without wind, based on the system specified by Eq. [5.8]?

15. Cost out a homemade system similar to that of Table 5.2 with the same power output, using generators producing 45 A at 500 revolutions per minute and selling for $30. Assume 18 batteries are required.

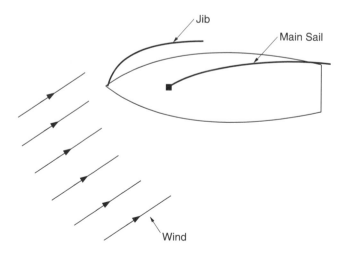

**Fig. 5.8. Sailboat with jib.**

*The following two projects require actual construction.*

16. Build a Savonius rotor and test it, using a generator for a bicycle light.

17. Build a full-scale sail wing or a model of one. Test it in the wind and estimate the speed at which it revolves.

18. Construct a propeller of wood and test it in the wind. Estimate the speed at which it revolves.

*Chapter*

# 6

· · · · · · · · · ·

# TOOLS

· · · · · · · · · · · · · · · · · · ·

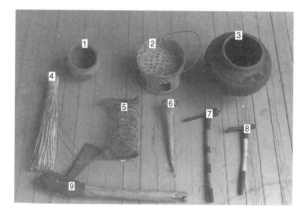

**Handmade tools used in central Africa. 1, Cooking pot; 2, stove; 3, storage pot; 4, broom; 5, fishtrap; 6, mousetrap; 7, hand hoe; 8, adze; 9, axe.**

## DEFINITION OF AN APPROPRIATE TOOL

The usual definition of a tool refers to an implement used for hitting, cutting, digging, and so on. The authors prefer one that is "a means, an aid to accomplishing a task." This definition is broader. It includes not only a hammer but also a book, a pen, or a guitar.

In Chapter 1 we described how E. F. Schumacher used the cost of a workplace as a gauge of the height of the technology used in that workplace. A $2 plow is the tool needed by a subsistence farmer. A $2 billion factory is the tool of someone fabricating integrated circuits. The inference is that the farmer is engaged in low technology and the circuit fabricator is practicing high technology. Schumacher pointed out that a middle ground exists where tools may cost, for example, $500. A worker using such tools practices intermediate or Appropriate Technology.

An often neglected point regarding tools, whether of high, intermediate, or low technology, is that they require skill on the part of the user. One must be able to grasp the "handle" and know how to make it perform. Like a lever, the tool is an extension of the user. It depends on the user to transfer her or his sense of quality and utility to the task being performed.

Some programs have given tools to unprepared workers, neglecting the problem of teaching potential users how to achieve quality results easily and efficiently. Tools in unskilled hands have laid waste to more material, more quickly, than one would care to imagine. Again, it is not enough to supply hardware. Software, in the form of instruction, is just as essential. In this chapter we will examine bicycles, pumps, and cookstoves as tools.

## BICYCLES

The usual response of most Americans to questions about bicycles is, "I had one when I was a kid" or "Sometimes I go for a ride in the country if the tires aren't flat," but to millions of the world's population, bicycles are the basic mode of transportation. Neal Pierce [5], a syndicated columnist, wrote the following of bicycles in China:

> *Chinese cities live and breathe by the bike. On a first visit in 1979, I was startled to hear city officials bemoaning the onslaught of cycles, asking how it might be "controlled." I thought, and still do, they were asking the wrong question. Compared to the autos that make motorized hells of so many world cities, the bicycle is light, compact, healthy for its riders, and environmentally benign. The people, animals and assorted cargoes the Chinese pile onto bicycles defy the imagination. Wives, babies, girlfriends, (boyfriends), grannies, hat racks, huge loads of hay, stacks of furniture swaying precariously from side to side — almost anybody and everything gets balanced on the handlebars or behind the rider.*

In most areas of the world the bicycle is the transportation of choice. Autos are too expensive and public transportation is overcrowded and under-funded. Figure 6.1 shows a bicycle in China.

**Fig. 6.1. Bicycle in China. (Reproduced with permission from *China Reconstructs* [8].)**

## *Bicycle Technology*

How is it that a person on a 30-lb bicycle can go faster and further in a day than a person walking, even though the walker's load is 30 lbs less? One reason is that rolling friction, predominant in a bicycle, is less than the sliding friction predominant in walking. Another reason is that the force and speed used in propelling a bicycle matches the force and speed at which human muscles are most effective. A final reason is that the frame and handlebars help support the cyclist's body and additional loads being carried, so more of the bicyclist's energy goes to producing motion compared to the walker.

Bicycles evolved to their current form about 100 years ago and most present-day designs are fairly standard, but it is instructive to consider some of the design decisions. In designing the frame the trade-off is between strength and weight. Ordinary bicycles in China and India, where big loads — approximately 200 pounds — are carried on bumpy roads, might weigh 40 lbs. A 10-speed bicycle used for pleasure in the United States might weigh 20 lbs. The frame can be improved by using expensive metal alloys, which are stronger for the same weight than the steel normally used in bicycles. Metal failure in a bicycle frame is usually produced by "fatigue"—repeated small bending rather than a one-time overload. Therefore, for durability, one would like a frame which would not bend, but a somewhat flexible frame is more comfortable than a rigid one. Frame design is in part a trade-off between durability and comfort.

The main source of friction, which makes pedaling difficult, is the tires against the ground. Another source is internal to the bicycle, at the wheel hubs. The friction from this source is small because ball bearings are used in the hubs. Early bicycles did not have these ball bearings and the resulting friction made starting difficult. Tire friction is increased if the tire pressure is low—force must be exerted constantly, as the wheel turns, to compress different parts of the tire; in a sense, one is always pedaling up hill. Although rubber tires create more friction than uncompressible ones, they soften the ride by absorbing bumps. Without rubber tires the ride would be uncomfortable and the frame would break sooner. Once the bicycle is moving normally the greatest source of friction, and the one that limits the top speed, is air resistance—pushing the air ahead of the bicycle out of the way. Very high-speed bicycles have a streamlined body to reduce air resistance—one bicycle so equipped attained a speed of 65.48 mph but this bicycle was hardly an intermediate technology device (it also had special tires and a lightweight frame).

Bicycles generally have gears and a chain. The purpose of the gears is to allow the cyclist's legs to move at their optimum rate, 50–100 rpm no matter how fast the bicycle wheels are moving. The chain, of course, transfers the power from the pedal to the wheels. A well-oiled, clean chain is about 98.5% efficient; that is, only about 1.5% of the power from the pedals is used to move the chain itself.

Brakes are, for obvious reasons, essential. On some racing bicycles one pushes backwards on the pedals to brake but the work required is greater than that of rubbing something against the wheel—the usual type of brake. The major problem with conventional brakes—rubber pieces pushed against the wheel rim—is that they lose effectiveness when wet. The braking action is produced by the friction of the rubber brake shoe against the metal rim and this friction nearly disappears when there is a thin film of water between the rubber and the metal. Stopping a bicycle on a rainy day can be perilous! So-called "coaster brakes" work by forcing metal rings inside the rear hub against each other. Coaster brakes are enclosed so they are not affected by weather but only work on the rear wheel. Enclosed drum brakes for front wheels have been developed but they require a heavier wheel and more force from the bicyclist.

## Maximum Speed, Power, and Capacity
## of a Bicycle

As previously mentioned, a sophisticated bicycle in special conditions has achieved a speed over 65 mph for 200 m. A more typical speed for an average bicyclist is 12 mph. This corresponds to about one-tenth of a horsepower or 75 W and can be sustained for at least an hour. The load capacity of a heavily built bicycle seems to be limited more by the balancing skill of a cyclist than by the strength of the bicycle; 300 or 400 lbs can be carried without damaging a strong bicycle.

## Uses for a Bicycle

We divide uses into two categories: transportation and stationary power. Because the technology of the bicycle itself is so mature, it seems that fruitful approaches to using bicycles will be mostly either in the supporting infrastructure — making bicycle transportation more convenient — or in imaginative ways of applying their power — driving different kinds of tools.

### Transportation Uses

Bicycles certainly have many advantages when used for transportation. In crowded cities, including those in the United States, a bicycle tends to be a faster vehicle for commuting than an automobile, at least up to distances of 5 miles, because of traffic. A bicycle goes nearly door-to-door and is easier to park. It does not pollute and can use narrower roads. It does not use gasoline. It is cheaper than an automobile to purchase and much cheaper to run. Bicycle maintenance can be done by more owners than can auto maintenance. Bicycle riders tend to enjoy riding and their health is probably improved by the exercise.

What are the drawbacks of a transportation system based on bicycles? The following is based on the book by Whittin and Wilson [6]:

- No protection from rain or snow
- Poor braking in wet weather
- Difficulties in seeing a moving bicycle at night
- Difficult to carry large packages or children
- Long or steep hills are tiring
- Many commuters will not want to cycle more than 5 miles to work

- Fairly frequent failure of key parts, particularly tires and brake and gear cables (It is worth noting the best known Chinese bicycle, Flying Pigeon, uses metal rods instead of brake cables)

Some of these problems could be resolved by covered, illuminated, and redesigned roads, which would certainly be cheaper and more environmentally benign than superhighways. One approach to using bicycles to carry large loads is the "Oxtrike," discussed in the next section.

### The Oxtrike

Cycle rickshaws are used often as taxis, in many countries in Asia (Fig. 6.2). They are tricycles, with the driver in front and the passengers, as many as three, seated over the rear wheels. Bicycle parts — wheels, hubs, handlebars, sprockets, etc — are often used, with a longer chain. The awning gives some protection from sun and rain but pieces of waterproof cloth or plastic sheeting are often carried to be wrapped around passengers in heavy rain. Cycle rickshaws can also be used to carry freight. Traditional rickshaws powered by a person running have been essentially replaced by these cycle rickshaws, perhaps because the latter go faster or perhaps because they are considered less degrading for the operator.

The usual cycle rickshaw suffers from two defects: They are hard to stop and hard to start, especially on a hill. The stopping problem results from using only front wheel brakes; it is difficult to modify standard bicycle parts to achieve rear-wheel braking. The starting problem also results from use of standard bicycle parts; the gear ratio is not proper for the much heavier loads in a cycle rickshaw.

**Fig. 6.2. Cycle rickshaw. (Reproduced with permission from *Appropriate Technology* [7].)**

The "Oxtrike" (Fig. 6.3), was developed at Oxford University with support from Oxfam. It is an improvement on the cycle rickshaw and is currently being tested in various countries. Its design includes both front and rear brakes, the latter is a novel design actuated by a separate pedal. A standard three-speed gearbox is used and is located midway between the pedals and the rear wheels, so two chains are necessary but they are standard. The frame is not made from thin-walled tubing, as in most bicycles, but rather standard steel sheets, folded and welded. This kind of construction can be done by local metalworkers, whereas thin-wall tubing requires special tools for cutting and welding. Various kinds of bodies, for carrying people or baggage, can be mounted on the frame.

One reason we mention the Oxtrike is to show how a machine, a bicycle in this case, used in Europe or the United States can be modified to meet needs elsewhere. In this modification, attention was paid to fabrication; the Oxtrike was designed to be made using locally available techniques and only commonly available imported components were used.

Bicycles, like cars, can be used to haul loads. A "village ambulance" is shown in Fig. 6.4. This was made in Malawi and uses standard bicycle wheels and other parts. The same workshop also produced wheelchairs, again using standard bicycle parts.

### Stationary Power

Power generated by stationary pedaling can be used for many things — electric generators, water pumps, small tools such as grinders, drills, or kitchen appliances. The major design question is whether one uses an ordinary bicycle, temporarily mounted to run a machine, or whether one

**Fig. 6.3. Oxtrike. (Reproduced with permission from *Appropriate Technology* [7].)**

**Fig. 6.4. Village ambulance.**

builds from scratch using bicycle parts. In a sense, the bicycle exercise machine used in many homes is an example of the latter, but the power is wasted.

An example of a machine which uses an ordinary bicycle is the grain mill shown in Fig. 6.5; the grain is actually milled in the box in the right side of the figure. In Fig. 6.6 a water pump driven by a bicycle is shown.

**Fig. 6.5. Bicycle-powered grain mill. (Reproduced with permission from**
***Pedal Power* [4].)**

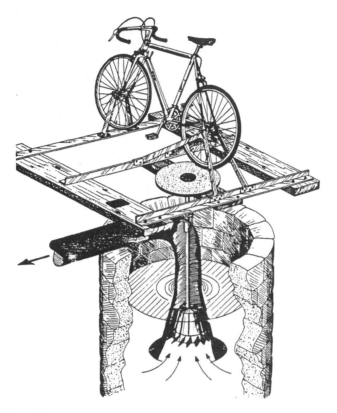

**Fig. 6.6. Axial water pump powered by a bicycle. (Reproduced with permission from *Pedal Power* [4].)**

Adapters allowing conventional bicycles to power machines are fairly easy to build. They must be strong and rigid because the forces are considerable. In these adapters power is usually taken from the bicycle by a rubber-tired wheel, a roller, or a pulley firmly pushed against the rear tire. Electric power can be generated using an automobile generator. One could also use the small generators sold to power bicycle lights, but these produce much less current. In the authors' courses, bicycle-powered electric generators as well as a device to split logs have been built by students.

Instead of adapting an existing bicycle, a specially built pedal-powered machine can be designed. In Fig. 6.7, a home-built frame, seat, set of pedals, and a chain with sprocket are combined to give a power source. This particular model was built at the Rodale Institute, Emmaus, Pennsylvania.

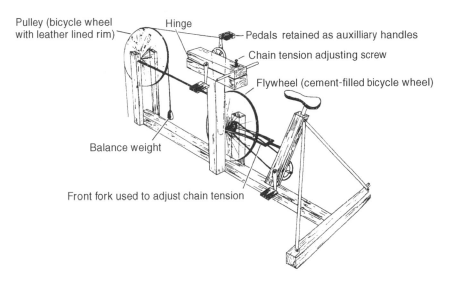

Pulley (bicycle wheel with leather lined rim)

Hinge

Pedals retained as auxilliary handles

Chain tension adjusting screw

Flywheel (cement-filled bicycle wheel)

Balance weight

Front fork used to adjust chain tension

**Fig. 6.7. Power machine. (Reproduced with permission from *Pedal Power* [4].)**

It is used to drive hand tools — grinding was especially effective — and kitchen appliances. One useful application is to set the machine up as a winch at the end of a garden row. The machine pulls small farm machines across the garden, an alternative to using a tractor, which compacts the soil and can only be used if the rows are far apart. Building instructions for the device are available from the Rodale Institute — costs depend on the builder's skill with tools and with scrounging parts (the address is provided in Ref. [4]).

## *Future Uses of Bicycle-Related Technology*

Bicycles seem to fit our definition of an appropriate technology. They certainly can serve a useful function. They are easy to maintain. Many of the parts can be repaired by the owner and the others are widely available. In use, they are labor rather than capital intensive and are easy on the environment. Bicycles do have disadvantages. For some applications — carrying big loads or wet-weather trips — they are inconvenient. In some parts of the world they are looked down upon as backward. Manufacturing some of their components, such as the hubs which contain ball bearings, requires fairly sophisticated machinery. Will they be used more in the future? It is simply not clear.

We have mentioned only briefly "hi-tech" embodiments of pedal power. Besides the high-speed bicycle which was mentioned, human-powered aircraft and boats are being developed. In 1979 the aircraft "Gossamer Albatross," entirely human powered (the pilot pedaled to turn the propeller), crossed the English Channel. A human-powered boat has achieved 13 knots over a short distance. The operator of the boat pedals, thus turning a propeller. The craft is equipped with hydrofoils and lifts out of the water. Such machines are fun to experiment with and lessons learned in their development can be useful elsewhere.

# PUMPS

Water pumping can greatly increase food production. Many people who live where the climate has a pronounced wet season followed by a pronounced dry season grow rice extensively, but only in the wet season because rice cultivation requires much water. Often, nearly all the arable land is in use. Adding some type of irrigation system allows a second crop to be grown in the dry season, thus effectively doubling the available land.

To demonstrate how water pumps have been introduced, a specific example will be considered. In Bangladesh, a dry-season wheat crop can complement the wet-season rice. To flourish, a wheat crop needs at least 10 cm of irrigation water during the growing season. This means 100 liters per square meter must be supplied or 100,000 liters per 1000 square meters (1000 square meters is a small field). In 1980 the only irrigation available to a small farmer was the lifting of buckets from a dug well. For a person lifting 1 liter per second, at least 29 hr of hard labor are required to irrigate a small field.

Many agencies attempted to alleviate the pumping problem with a variety of solutions; diaphragm pumps, bicycle power, twin cylinders, and other imported technologies were tried. Each eventually rusted away in disuse. Farmers were hard to impress, probably because they had seen other new devices fail in practice, and continued to rely on their traditional bucket or swing basket irrigation.

In 1976, the Rangpur-Dinajpur Rehabilitation Service (RDRS), operated by the Lutheran World Federation, began working on small irrigation pumps. Their initial offering of a pedal-powered, flywheel-assisted, one-cylinder plunger pump gained some use. However, it was too expensive and raised only a little water. (This pump is discussed from the policy viewpoint in Chapter 17.)

After several more false starts, RDRS designed and built a pump using treadles to take advantage of the operator's legs and full body weight. One of the treadle pump's developers relates the farmer's reaction:

> *When the following day the first treadle pump was assembled, mostly from components at the workshop, and installed among other pumps on the roadside, it immediately attracted passing farmers in a way other pumps had not. They sized up the ease of operation, the output, and the price tag and demanded their own specimen. (Personal communication, Javed Ahmed, Sept. 20, 1987.)*

The twin treadle pump is illustrated in Fig. 6.8. The prime mover is a person pedaling. The pump is actually a pair of pumps very similar to the standard displacement pump, which is described later in this chapter. Twin cylinders, operated alternately, provide a steady discharge, as opposed to a single cylinder which discharges only on the upstroke. By basing the treadle pump on a familiar and widely distributed displacement pump the developers have ensured that the user will be acquainted with its internal workings and that spare parts will be easily available.

A practiced treadle pump operator can work a 2-hr stint, producing 2 or 3 liters per second at 2–4 m lift. (The lift or head is the distance from the water level to the discharge spout.) Most of the materials necessary to construct the pump were locally available. The frame, treadles, and pipes

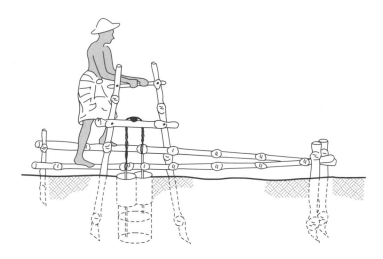

**Fig. 6.8. Twin treadle pump. (Reproduced with permission from**
***Appropriate Technology* [9].)**

can all be made of bamboo. The cylinders are made from sheet metal, PVC tube, cast iron, or concrete. The imported items — axles, seals, pistons, and rods — are readily available because they are replacement parts for widely used pitcher pumps.

Because twin treadle pumps are fairly fragile and have a life of 4–6 years, craftspeople in each community were trained to do repairs and installations. A good team of two or three people can make the bamboo pipes and strainers and frame and sink and install a pump in a day.

The cost of a complete pump installation is approximately $15–20, which is within reach of all but the poorest farmers. As of April 1985, almost 40,000 treadle pumps had been sold. The technology is currently being transferred to China, the Philippines, Mauritania, and Tanzania. The treadle pump was recently nominated as one of the 100 best innovations for development [9].

## Usefulness of Pumps

Pumps supply three basic water needs:

- Domestic — providing water for household use; drinking, cooking, cleaning, and bathing
- Livestock — drinking water
- Irrigation — giving farmers better control over their water resources

The value of the water and therefore the justifiable cost for pumping water is dependent on the usage. Obviously, domestic consumption is given the highest value. Water for livestock is a close second because a relatively small quantity of water can sustain a large and valuable herd. Irrigation use has the lowest value. Large quantities of water are needed to produce crops of modest worth; for example, to produce 5000 kg of corn, 12,000 cubic meters of water is needed.

Many types of power sources can be adapted to driving pumps. Human-powered pumps are probably the most prevalent worldwide. They can provide a moderate supply of fresh water on demand, are well understood, and are easily implemented. If more water is required, animals are used to provide the muscle because they do not balk at pushing a pump all day.

When the materials and technology are available, wind, water, solar, electric, or biomass power sources can be employed. The choice of power source depends on the economic resources, water demand, water source, and the technical resources of the community.

## *Types of Pumps*

It is helpful to define two categories for pumps: lift (suction) type and force type. Lift pumps pull the water up from the well. Actually, it is the pressure of the atmosphere at the bottom of the well which pushes the water up. The pump really only removes water from the top of the pipe so more water can be pushed up. A lift pump is located at the top of the well and is limited to a lift of 6 or 7 m, corresponding to atmospheric pressure. Force pumps actually push the water up from the bottom of the pipe.

The most familiar and ubiquitous pump is probably the displacement pump—a lift pump. At the beginning of the twentieth century windmills drove millions of displacement pumps. When one thinks of early farms the image of people pushing up and down on the pump handle comes quickly to mind.

Figure 6.9 shows a cutaway of a typical displacement pump. It operates as follows: On the upstroke the piston valve (plunger bucket) closes and the foot valve opens. Atmospheric pressure pushes the water up the suction tube. On the downstroke, the foot valve flaps close and the piston valve opens, allowing the piston to pass through the water already in the cylinder. On the next upstroke more water is pushed up the suction tube and the water already in the cylinder is lifted and forced out the spout.

If the pump seals and valves are in good condition the pump does not need to be primed. Perhaps an explanation of priming is in order. When a pump is not in operation, the water in the cylinder and suction tube may leak back to the well so the valves do not seal. Adding water to initiate pumping is called priming. A well-designed displacement pump is capable of pumping the air out of its system and so does not need priming.

Force pumps are located at the water level—usually below ground level—and push the water up from the well. Actually, most lift pumps will function as force pumps if they are submerged, but of course they must be supplied with power from the surface. Examples of force pumps are centrifugal pumps, rotary pumps, jet pumps and chain pumps.

Centrifugal pumps operate by throwing water from the center of the pump to the outlet located on the outer casing. Figure 6.10 shows a centrifugal pump. At the center of the impeller there is a low-pressure area which draws more water up the suction tube. Initially water must be in this center area if low pressure is to be produced when the impeller rotates, so a dry pump must be primed. To avoid having to prime the pump before each use a check or foot valve is installed in the suction tube. This valve prevents water from flowing back into the well, i.e., it only allows flow in one direction. Centrifugal pumps have a 5- to 10-year lifetime, they require little maintenance, and in large installations they have an efficiency of 80% (with 50–70% in smaller ones).

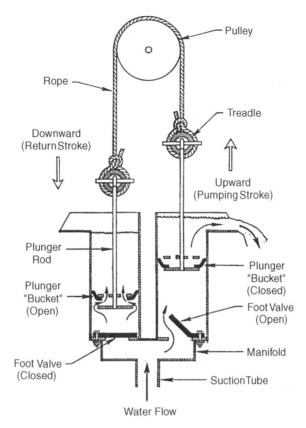

**Fig. 6.9. Displacement pump. (Reproduced with permission from IRRI [11].)**

There are several variations of axial or rotary pumps. The best known is the Archimedes screw, which is a broadly threaded screw rotating in a snugly fitting tube. Another axial pump is made by mounting a marine propeller in a pipe. When the propeller is driven it forces the water to the outlet. Figure 6.11 shows an axial pump. Axial pumps are useful in low-lift, high-volume applications. They are relatively insensitive to sediment in the water.

A jet pump (Fig. 6.12), is really two pumps. The secondary pump provides water to the primary pump. The primary pump has no moving parts; it is merely a nozzle whose outlet is a high-velocity stream inserted in a tube. The nozzle creates an area of low pressure which draws more water into the tube. Jet pumps are a good choice when the lift is more than 8 m or

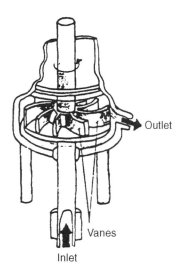

**Fig. 6.10. Centrifugal pump. (Reproduced with permission from *Renewable Energy Technologies* [12].)**

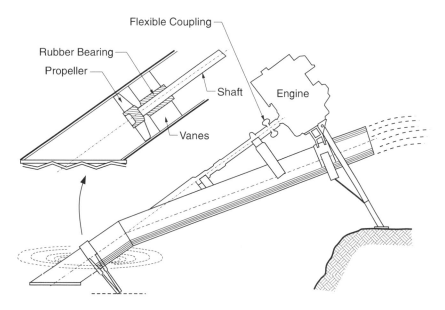

**Fig. 6.11. Axial pump. (Reproduced with permission from IRRI [11].)**

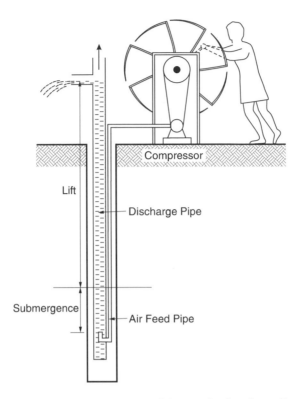

**Fig. 6.12. Jet pump. (Reproduced with permission from *Renewable Energy Technologies* [12].)**

the bore of the well is narrow. Their efficiency is somewhat less than that of a centrifugal pump.

The choice of a pump for a water supply system is based, like most appropriate design decisions, on local conditions, particularly needs and available resources. The purpose of the water (domestic, livestock, and/or irrigation), power sources, local customs, water source (shallow or deep), economic resources, climate, and local expertise all have an impact on the success of a water project. In a later chapter a pump is discussed which was not accepted in the Sudan because the users thought the operators would sit in an undignified position; use of the pump violated local custom.

The treadle pump in Bangladesh thrived because it responded to local needs. The cost is kept to a minimum, the pump is simple, and it has a sustainable and high output. Had the designers failed at any of these goals — cost, dependability, and output — the treadle pump would have joined its cousins rusting in a field.

# COOKSTOVES

In 1976, Guatemala was hit by devastating earthquakes. Many lives were lost, many homes were destroyed, and the region's way of life was severely disrupted. The high demand for wood for rebuilding caused a shortage of fuelwood. To deal with the shortage a program was started to improve stove efficiencies.

The stove implemented in Guatemala, a version of the Lorena stove (*lodo* = mud, *arena* = sand) was developed by the Approvecho Institute in Eugene, Oregon [17], [18]. The stove was constructed from a solid block of sand and clay into which were cut a firebox and three holes for pots. A metal damper and chimney were incorporated to complete the stove as shown in Fig. 6.13. The stove was well received; it reduced smoke and raised the cooking level off the floor. Through 1983 about 6000 Lorena stoves were built. The major problem with the stove is that the material tends to crumble, thus limiting its life to about 1 year. Other problems do exist. Because it has a large mass the stove takes a long time to heat up. The chimney increases the cost of the stove to a point that it could not be built without a subsidy. To understand why this stove design and other designs emerged as they did, it is necessary to understand how a stove works.

## *Stove Technology*

For combustion to take place, the following are needed: fuel, air (oxygen) to react with the fuel, and an ignition source to initiate the reaction started. A match is a good example of a combination of these components. The heat for ignition is generated by friction when the match is rubbed across the striker. This heat ignites the chemicals in the match head, which in turn lights the body of the match. Once combustion is started it will continue until the fuel (the wood) or the air is consumed.

The task facing stove designers is how to control the combustion and how to use the heat generated most efficiently. We will consider control first. An open fire is controlled by the loading of fuel and the whim of the wind. In a strong breeze a fire will consume its fuel more quickly and give off more heat in a given amount of time. The reason the rate of combustion depends on the wind is that hot air rises, so a fire creates an updraft, pulling air in from the side and pushing it out above the flame. Wind increases the amount of air flowing into the fire.

A pot supported by three stones with a fire beneath is the most prevalent open-fire construction. If the fire is built on the ground, the fuel closest to the ground is starved for air and does not burn completely. Because insufficient air creates incomplete combustion, these fireplaces

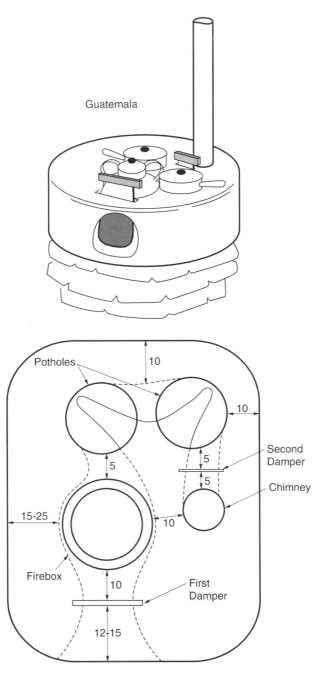

**Fig. 6.13. Lorena stove from Guatemala. (Reproduced with permission from *Cookstove News* [17].)**

tend to be smoky, creating health and hygiene problems, although smoke does keep insects away. The efficiency of an open fire (heat that ends up in the pot divided by heat potential of the fuel × 100) is 10–15%. Three-stone fireplaces cost nothing and can be built nearly anywhere. They provide heat, light, and a social place, so they do have advantages.

The first tack for improving efficiency is to control the flow of air around the fire. One method is to raise the fire off the ground by building it on a grate. Then the air will flow from beneath the fire which ensures complete combustion of the fuel. To gain greater control one can build the fire in a box with a damper to control the air intake and a flue to vent the exhaust gases out of the box.

By controlling airflow one ensures complete combustion and can control the rate of fuel consumption — less draft will slow the fire and more draft will speed it up. When things burn in air, they do so at a fairly constant, fuel-dependent temperature. What can be changed is the rate of consumption and therefore the amount of heat liberated in a given length of time. Fires do not so much burn hotter or colder as they do faster or slower. The faster a fire burns the "hotter" it appears to be.

A typical burn demonstrates the different ways a fire is controlled. A fire begins in a fuel-limited condition until complete ignition is achieved. The fire is then draft limited, that is, controlled by the air until the fuel is depleted enough to affect the rate. The fire continues to be fuel limited until it exhausts the fuel supply.

Heat is transferred by three mechanisms: conduction, convection, and radiation. These mechanisms are described in more detail in Chapter 12. The problem in designing a stove is to maximize the energy transferred to the cooking pot and minimize that wasted. Heat can be radiated directly from the flames to the pot. It can also be carried (convection) by the hot gases to the pot. In both cases, the designer should do all that is possible to increase the effectiveness of the heat transfer, for example, by directly exposing the pots to the flame or by ensuring the hot gases cover the bottom and sides of the pots. Unburned fuel between the fire and the pot reduces the amount of heat transferred, as does direct passage of hot gases up the chimney.

Efficiency, defined previously as the ratio between useful heat produced and potential heat in the fuel, increases if losses are reduced. Losses result from incomplete fuel combustion and continued burning after cooking is completed. They also result from heat escaping through stove walls, cooking vessels, the chimney, and uncovered spaces. Of course, some of this lost heat is useful in warming the room.

As previously mentioned, an open fire smokes. The smoke can be unhealthy for the cook and, especially in a closed room, for other members of the family, producing both eye irritation and lung difficulties. On the other hand, the smoke can keep insects away and even small animals (e.g., lizards in a thatch roof). The smoke might also be useful for curing meat hung high in a room.

## *Stove Construction*

There are three components that can be arranged to make a stove: the firebox, the damper, and the flue/chimney. The other parameter we control is the material out of which the stove is constructed. Typically stoves are categorized as low mass or high mass. This refers not to their weight but to their thermal mass. A high-mass stove will heat slowly and hold the heat for a long time. High-mass stoves are constructed of mud clay, sand, bricks, or rocks — materials usually present in rural areas. Low-mass stoves heat and cool quickly. They can be made of metal or fired clay. Metal stoves are often made from scrap metal, such as discarded oil drums and paint cans.

It is easy to see that each type of stove has its strong points and drawbacks. The high-mass stove is probably the simplest to construct, it retains heat well, and it is helpful for space heating as well as for cooking. Most of the high-mass materials are fairly brittle, however, and tend to crumble under use. Low-mass stoves are simple and portable. A low-mass stove in the courtyard of a Zambian house is shown in Fig. 6.14. They are used primarily in urban areas for outdoor cooking. Their drawbacks include being unstable under heavy loads, requiring fuel to be cut into small pieces, and losing much heat through radiation.

**Fig. 6.14. Low-mass stove.**

The designers in Guatemala had the considerations above in mind when they produced the Lorena stove [17] (Fig. 6.13). Its high mass makes it useful for space heating in addition to cooking. It has an efficiency of 20–25%. Although this is better than an open fire, it may not be sufficient to justify the cost ($20), mostly for the chimney. One advantage of the Lorena stove is that it is large enough to serve as a work area as well as for cooking.

The operation of the Lorena stove is straightforward. The two dampers control the flow of air. The sharp bends in the tunnel create turbulence under the pots, which ensures even heating. The dampers are simply small pieces of sheet metal, with handles, fitting tightly into grooves which hold them in position. The chimney is rolled metal and is the most expensive part. The stove is made by molding wet clay, building the stove in stages. Construction takes 3 or 4 days. Homemade tools can be used, such as a sifting screen, a shovel, a bucket, and mason's trowel.

A contrasting example to the Lorena stove is in the Damru Chulha designed at the Agricultural Tools Research Center, in Bardoli, India [13] (Fig. 6.15). The Damru Chulha is simple enough to be built by a local potter at a cost of about $4. Combustion is improved by the iron grate onto which the fuel is loaded. A clay ring, the "heat shield," is used to reduce heat loss from the cooking vessel. The Damru Chulha is about 30% efficient. Of course, this stove does not warm the room effectively and cannot be built by the homeowner.

A final stove is shown in Fig. 6.16 — a rim oven from Botswana. The stove is made from two truck rims welded together and surrounded by mud

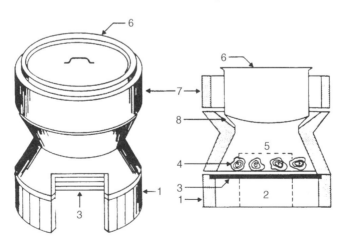

**Fig. 6.15. Damro Chulha stove. 1, ash pit construction; 2, ash pit window; 3, grate; 4, fuelwood; 5, fuel-charging window; 6, cooking vessel; 7, heat shield; 8, three ribs to support vessel. (Reproduced with permission from Agricultural Tools Research Centre [13].)**

**Fig. 6.16. A rim oven. (Reproduced with permission from Rural Industries Innovation Centre [19].)**

and bricks to maximize fuel efficiency. It is especially meant for small-scale commercial bread baking and has been successful in developing small businesses in rural areas [19].

Stove technology has changed rapidly over the past 20 years. Efficiencies of 35–40% are possible, but it must be recognized that the stoves must be in harmony with local custom to gain acceptance [14, 15]. Improved designs can serve as a valuable tool for teaching users what makes a stove efficient and thus allowing them to create the next generation of stoves [16]. In Chapter 16, it will be shown that improved domestic stoves can make a major improvement in a nation's use of energy.

# REFERENCES

## *Bicycles*

1. Abbott, A. V., Brooks, A. N., and Wilson, D. G. (1986, December). Human-powered watercraft. *Sci. Am.* **255**(6), 120–130.

   *Describes the pedal-powered hydrofoil driven by a propeller.*

2. Drela, M., and Langford, J. S. (1985). Human-powered flight. *Sci. Am.* **253**(5), 144–151.

   *Describes development since the "Gossamer Albatross," aimed at faster, more versatile aircraft.*

3. Hoda, M. M. (1984, December). Appropriate methods for improvement of cycle-rickshaws. *GATE* **84**(4), 34–36.

   *The author, director of an Appropriate Technology group in India, describes generally the cycle rickshaw and what it needs.*

4. McCullagh, J. C. (1977). *Pedal Power*. Rodale Press, Emmaus, PA.

   *Good source of ideas and plans; includes much more than bicycles — all sorts of treadle and foot-driven machines, easy to read.*

5. Pierce, N. (1987, March 23). China's bikers, how they increase. *Providence Journal*. A12.

   *Bicycles in Shanghai.*

6. Whittin, F. R., and Wilson, D. G. (1982). *Bicycling Science, Second Edition*. MIT Press, Cambridge.

   *Good solid science, for example, starts with a discussion of the efficiency of human muscles. The fundamental limitations of bicycle power are discussed very clearly, as are modern refinements. A good history of bicycles and several peeks into the future.*

7. Wilson, S. S. (1977, February). The oxtrike. *Appropriate Technology* **3**(4), 21–22.

   *Gives design details of the second model built.*

8.  Youma, Z. (1987, May). Ode to the bicycle. *China Reconstructs*, 34–39.

    *What it is like to own a bicycle in China. The pictures are much fun, and so is the article.*

## Pumps

9.  Innovations for development. (1990, December). *Appropriate Technology* **17**(3), 29.

    *Short article describing RDRS pump and award.*

10. Appropriate Technology International (1989, May). Treadle Pumps in Cameroon and Mali, Bulletin 18.

    *Description, construction and commercial appraisal.*

11. IRRI, P. O. Box 935, Manila, "MAF_IRRI Industrial Extension Program for Small Farm Equipment—Annex for Annual Report (1984–1985)," Agricultural Engineering Department. Also "Agricultural Mechanization in ASIA, Africa, and Latin America."

    *IRRI stands for International Rice Research Institute. They have done much with irrigation pumps, as well as high yielding seed varieties.*

12. Kristoferson, L. A., and Bokalders, V. (1986). *Renewable Energy Technologies*. Pergamon, Oxford, UK.

    *Presents good overview papers on a variety of technologies, including pumps and stoves.*

13. Agricultural Tools Research Center, Suruchi Campus, Post Box 4, Bardoli 394601, Gujarat, India.

## Stoves

14. Barnes, D. F., *et al.* (1944). What makes people cook with improved biomass stoves; A comparative international review of stove programs (World Bank Technical Paper No. 242, Energy Series). The World Bank, Washington, DC.

    *Reasons for using improved stoves and how to encourage people to use them; much information.*

15. Campbell, J. R., and Bezuayenae, S. (1991, March). Improved stove in urban Ethiopia. *Appropriate Technology* **17**(4), 29–31.

    *Describes the stoves but mostly how the project succeeded.*

16. Clarke, R. (1985). *Wood-Stove Dissemination*. IT Publications, 9 Kind Street, London WCZE 8HW, UK.

    *The conclusions of many different wood stove projects; practical and easy reading.*

17. "Cookstove News," Aprovecho Institute, 442 Monroe, Eugene, Oregon 97402.

18. Evans, I., and Boutette, M. (1981). *Lorena Stoves*. Appropriate Technology Project of Volunteers in Asia, Stanford, CA.

*On designing, building, and testing wood-conserving cookstoves; much practical information.*

19. Rural Industries Innovation Centre, Private Bag 11 Kanye, Botswana, Catalogue of Goods and Services, 2nd Edition, page 23, 1992.

*The Centre has designed many useful devices beyond this rim oven. A newer version of the catalogue is available.*

# PROBLEMS AND PROJECTS

## *Bicycles*

1. If you wanted to build your own bicycle, how would you do it? What parts would you buy? What parts would you make yourself, perhaps after redesigning?

2. Why might a foot pedal-powered machine give more power than one operated by the hands?

3. A person using pedals can produce one-tenth of a horsepower for a long time and perhaps twice that for several minutes. Is such an amount of power of any real use in the home or on a farm?

4. Consider each of the drawbacks to a bicycle transportation system and give ways of overcoming them.

   *The following projects require actual construction.*

5. Design and build a tricycle, using as many standard bicycle parts as possible, to carry four small children, with one adult doing the pedaling.

6. Design and build a machine, in which an ordinary bicycle can be mounted, to produce electric power.

7. Design and build from standard bicycle parts (as much as possible) a machine to power a circular saw for a home workshop.

8. Design a modification of a bicycle exercise machine that will power a small color TV set requiring 50 W.

## *Pumps*

1. Consider a particular situation, such as a rural, isolated farm in the United States, and chose an appropriate pump. Estimate the cost.

2. Describe how you might go about assessing the water needs of a community.

3. Consider a rural vacation home in the United States which gets its water from a well. How would you choose the kind of water pump?

*The following projects require actual construction.*

4. Design and build a bicycle-powered pump.

5. Consider a specific situation, such as a rural Central American village, and design a water system for it; build a model.

6. Establish a set of design criteria for a pump. Then make several designs meeting the criteria and then build the pump corresponding to the optimum design. This project might be done by several competing teams; resulting in a competition to build the most efficient or the cheapest pump that meets a set of design criteria.

## *Stoves*

1. What techniques might be used to measure stove performance?

2. Estimate the building cost of a homemade stove. If fuel costs $300 a year, how long until that stove pays for itself?

3. Of what locally available materials could you build a Lorena Stove in the United States?

4. How might you build a chimney for a homemade stove of local material?

5. (Project to be built) Build and test a low-mass stove designed to accommodate a specific culture/cooking style. Before making the design, state the culture/cooking style.

# *7*

· · · · · · · · · ·

# PHOTOVOLTAIC DEVICES

· · · · · · · · · · · · · · · · · · ·

## RURAL ELECTRICITY

Prior to 1980, a livestock farmer in Fredricksburg, Texas, relied on wind power to pump water for the farm's cattle. Wind power worked during the winter but in the summer, when the cattle needed water most, the wind seldom blew. The cost of installing electric power lines was $16,000. In addition to the capital cost of the electric lines, the electricity would have to be paid for as used—about $150/year. Instead of using the utility the farmer installed a photovoltaic (PV) system, costing $12,000, which powered a one-half-horsepower motor, sufficient to pump the needed water [5]. The only operating cost of the system is maintenance on the pump. The photovoltaic system was clearly less expensive than installing power lines in 1980, and since then the cost of PV systems has been reduced threefold, bringing the system cost to $4,000.

Photovoltaic devices complement wind-powered generators well; often, the sun is strongest during the times the wind is slowest. Using both ensures energy availability year-round — showing the advantage of using diverse sources of energy.

# SOLAR CELLS

Photovoltaic devices, also known as solar cells, produce electric current when light shines on them. They have no moving parts so their lifetime is 20 years or more. No maintenance of the cell is required, although the surface on which the light shines must be kept clean. As the reader probably knows, solar cells are used to power electronic calculators and other small electronic devices. A problem in many applications is the high cost. Solar cells tend to be used when the alternatives are even more costly or inconvenient. In Africa they have been used to power refrigerators in remote mission hospitals. In many parts of the world, including the United States, they are used extensively to power remote microwave relay stations where long-distance telephone signals are received, amplified, and retransmitted. The frontispiece to this chapter shows such a relay station. Much of the background for use of photovoltaics is provided in Chapters 3 and 12.

## *Principle of Operation*

Detailed understanding of what occurs inside a solar cell requires much physics but basic insight can be gained from simple ideas [2]. Most solar cells currently in use are made of silicon — a very common element. (Sand is silicon dioxide.) One reason that solar cells are expensive is that the silicon must be very pure — <1 impurity atom per 1,000,000 silicon atoms. Silicon, like every material, consists of atoms and the atoms contain electrons (small charged particles). When light shines on silicon, the electrons gain energy from the light and become free of the atoms. These free electrons can move. Their motion is current flow. Silicon is a semiconductor because it only conducts current when illuminated, or when electrons are liberated from the atoms in other ways, such as heat. A brief description of what occurs inside a solar cell is given in the Appendix.

The problem with making a solar cell is devising a way to collect all the free electrons so they will flow in the same direction in an external circuit. The method for accomplishing the collection employs small quantities of impurities. Usually these impurities are added to separate layers of silicon. Different kinds of impurities are mixed with the silicon. The effect of one type of impurity is to give one layer a small positive charge. The other type gives the other layer a slight negative charge, so a small voltage is produced between one layer and the other. Any electrons produced in the more negative material are attracted to the other, more positive, layer. Thin metal strips are inserted on this layer to collect the electrons so they can flow through a load and back to the first layer; these strips are visible on PV devices used in calculators. The number of impurity atoms introduced is very small—about 1 impurity atom to 1,000,000 silicon atoms.

Photovoltaic devices are expensive because they are made of thin layers of closely controlled material, deposited over large areas. A great deal of effort is being expended to make the process less expensive. Most of this effort is currently taking place in Japan, with the United States following suit. Maintaining tolerances, both on the purity of the base material and the placement of impurity atoms, requires highly trained people and complex equipment. The cost of devices will decrease as manufacturing experience accumulates. Manufacturers, however, will only get experience as more solar cells are sold. Support for PV energy, which encourages increased use, would bring prices down and the lower prices would increase sales, reducing prices further.

Promising areas of research to reduce costs, besides manufacturing techniques, include new materials which better match the sun's energy. Another research area concerns a way to stack cells on top of each other, so light not absorbed by one cell is picked up by one underneath.

Solar cells, of course, are made from the same materials as computer chips or other solid-state devices. The fabrication problems are different though. In a computer chip the concern is for precise placement of a pattern of impurities in a very small area. (The active part of a chip is much smaller than the chip itself, which must be big to have room for the pins.) Cost is not a primary concern in making a chip because the chip cost is a small portion of the overall equipment cost. In a PV device cost is a major concern, as are the large areas of semiconductor required.

Photovoltaics work well with little maintenance because nothing can really go wrong in a functioning cell. Of course, in fabrication a connection may not be made well or the layers may not be correct, but such defects appear immediately. A working cell will work a very long time   on the order of 20 years. The only problem is heat. When a cell is hot, extra electrons are produced and these electrons can overcome the voltage between the layers, reducing the efficiency of the cells. Cell temperatures should be kept below about 140°F. The cell is not, however, permanently damaged by overheating unless the material actually melts, which occurs at temperatures well over 300°F.

A typical solar cell generates 0.56 V. This value depends on the material used and is not adjustable. The amount of current produced depends on the area of the cell and the brightness of the sun. In full sun, a solar cell will produce about 0.28 A per square inch of cell; however, 20% of the voltage and current is lost internally so 0.45 V and 0.224 A are actually generated from a square-inch cell in maximum sunlight. Maximum sunlight means no clouds and also that the cell faces the sun directly—the sun's rays are perpendicular to the cell's surface.

## Panels of Solar Cells

Solar cells can be purchased individually or mounted on panels. On the panels the cells can be connected in series or parallel or both (Fig. 7.1). If the cells are connected in series the total voltage of several cells in series is the sum of the voltages of each cell. In a parallel connection the currents add—a parallel connection of several cells is exactly the same as one cell with a larger area, so the total current from a parallel combination of three cells is three times the current from each.

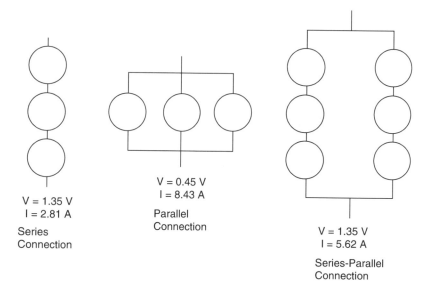

V = 1.35 V
I = 2.81 A

Series
Connection

V = 0.45 V
I = 8.43 A

Parallel
Connection

V = 1.35 V
I = 5.62 A

Series-Parallel
Connection

**Fig. 7.1. Connections of solar cells.**

A typical single cell sold commercially is close to circular in shape, about 4 in. in diameter. The output current in full and direct sunlight is about 2.8 A and the output power is about 1.26 W. The values for output current shown in Fig. 7.1 assume that each cell is this standard size — circular with 2-in. radius.

There are many different sizes and configurations of commercial panels. A typical power module which nominally produces 40 W and sells for $245 from Astro Power (30 Lovett Avenue, Newark, DE 19711) is shown in Fig. 7.2. This module produces a maximum of 17 volts and 2.80 amperes.

Solar cells are fragile: They are thin to reduce the amount of material required. It is therefore best not to handle them unnecessarily, therefore, unless one needs only a single cell for an application, purchase of an entire panel makes sense. In one interesting application in which single cells are deployed, heat from the cells — produced by the solar energy not converted to electricity — is used for heating a home. In this case, each cell was mounted on its own heat exchanger conducting heat from the cell to the heating system [3].

Another electronic device, which is actually very similar to a solar cell, is a diode. Diodes allow current to flow in only one direction. A diode is shown in the system diagram (Fig. 7.3). Current flows, as one might expect, from left to right in the diode but not from right to left. The reason for this will be given shortly. Diodes are also discussed in the Appendix.

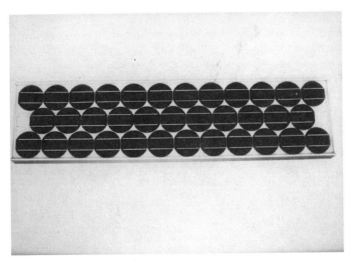

**Fig. 7.2. Commercial PV panel.**

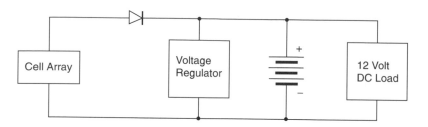

**Fig. 7.3. Solar panel with battery.**

## *Electric Systems Using Solar Cells*

Once one has purchased the solar cells and, if necessary, connected them together, not much more must be done, but it is necessary to consider cooling. Although the silicon will not be damaged unless the temperature is very high, the adhesive holding the cells on the panel will melt at a lower temperature, as will the solder. The effective maximum temperature is about 250°F, so the design must ensure that the cell temperature is less than this.

A possible systems problem is that no voltage is produced when the cell is not illuminated. Most PV systems include batteries for storage as shown in Fig. 7.3. When the cell is not illuminated (e.g., nighttime) the lack of voltage across the photocells would mean current could flow into them. The diode in Fig. 7.3 prevents the battery current from draining back into the solar cells when they are dark. A simple voltage regulator might also be included in the circuit to prevent overcharging the battery.

Another storage mode besides batteries, as discussed in Chapters 3 and 4, is to sell electricity back to the utility company. In this case, an inverter is needed to convert DC to AC; but a battery is not used. In some applications, such as refrigeration or irrigation, no storage is needed.

As in previous designs a choice of voltage levels and of AC versus DC must be made. The voltage available from a solar cell is DC and if we are using automobile-type batteries we would probably use 12-V as our output voltage. In this case, of course, we could only use 12-V DC appliances. If we prefer to use 120 V to reduce line losses, we could connect many (267 to be precise) solar cells in series. The 120-V DC could be stored by connecting 10 12-V batteries in series. Such a system would have lower line losses than a 12-V system, but many motors only work on AC so an inverter would probably still be necessary. Most commercially available inverters contain transformers and are designed for 12-V DC input and give 120-V AC output.

The cells are only part of the cost of a whole installation. The additional costs come from the panel and its mounting, storage, and the inverter. A very approximate estimate is that these so-called "balance of system costs"—the additional costs beyond the cells themselves—are equal to the cell costs. Incidentally, another term sometimes used is "power conditioner," referring to the inverter, which puts the voltage and current in a condition—proper magnitude, frequency, and wave shape—to be sold to the utility.

In Chapter 12, we will consider the optimum positioning of a solar heating panel. The same considerations apply regarding the angle of a panel relative to the sun for both PV and solar heating panels. Reflectors and lenses can be used to concentrate the sun's rays. A Fresnel lens, which is thin, is particularly useful because it absorbs very little of the sun's energy. Some systems use elaborate tracking devices so the panels always face the sun squarely. Some less elaborate systems are designed to be redirected manually once or twice a day to face the sun more directly. We will not pursue such refinements.

## SYSTEM DESIGN

The most important question in designing a system is to estimate the area of cells needed, that is, the number of cells. We continue with our example used in Chapter 3. We need 960 W for 5 hours. A typical commercial cell, 4 in. in diameter, has an area of 12.56 in.$^2$. How many cells are needed? We know we need 4800 W-hr and we know the power from a 4-in. cell, when the sun is shining directly on it (i.e., at solar noon), is 1.26 W.

What makes the calculation difficult is that if we direct the panel to get maximum energy at noontime, the rest of the time the panel will receive radiation at an angle, which is less effective. During the night, of course, we get no solar radiation at all. So how do we account in our calculations for the change in solar radiation during the day? Fortunately, people have worked out the effect of glancing radiation and lack of radiation at night. In a representative situation the average effective radiation power received during a day is 0.2 times the power of the peak radiation. Peak radiation is the radiation received at the optimum time of day (solar noon). Energy is power received multiplied by the length of time it was received. Therefore, if we multiply the maximum radiation received by 0.2 and by 24 hours we derive the same total energy as we would by adding the energies received at each instant of time:

$$E = P_{peak} \times 0.2 \times 24 \text{ W-hr} \qquad [7.1]$$

where $E$ is the energy received per day and $P_{peak}$ is the peak power from the cell power at noon on a cloudless day).

Peak power from one cell is 1.26 W and the energy desired from $N$ cells is 4800 W-hr, so

$$4800 = 1.26 \times N \times 0.2 \times 24 \qquad [7.2]$$

$$N = 794 \text{ (number of cells needed)} \qquad [7.3]$$

Typical cells are circular, 4 in. in diameter, so nine cells fit into a square foot. The area of cells is 89 ft$^2$ perhaps 6 × 15 (smaller than most roofs).

The cost of PV devices is usually given in dollars per peak watt. In this case, the number of peak watts from the system is the peak watt from an individual cell (1.26 W) times the number of cells, or

$$\text{Peak watts} = 1.26 \times N = 1000.4 \qquad [7.4]$$

A representative cost for cells in 1990 was $3 a peak watt (in 1978 it was $9). Therefore, our cost for the cells would be about $3000. If the balance of systems costs were equal to the cost of the cells then the total cost would be $6000. For a 20-year system using money borrowed at 12%, the annual costs would be $803.28.

## WHEN ARE PHOTOVOLTAICS APPROPRIATE?

Photovoltaics differ from other devices discussed in this book because they cannot be constructed by an amateur or even a good machinist. High-technology equipment is needed to purify the silicon and place the impurities in the correct places. On the other hand, PV systems are easy to put together from purchased panels and are certainly useful and environmentally benign. They make sense when the alternatives, such as power lines or batteries in calculators, are expensive or a nuisance.

Widespread use of PV devices would have significant advantages to society. Air pollution would be reduced because nothing would be burned to create power. (Of course, energy is used to manufacture the cells, but this is a one-time cost.) Electric power could be generated by solar cells to be used at that location — enough area exists on the roofs of most single-family homes to hold sufficient solar cells to meet the family's electric needs. Putting the generator close to the load means no losses in transmission and no land wasted for transmission lines. Use of PV power would give the homeowner control over her/his power system. Photovoltaics seem to be an ideal source — reliable, quiet, pollution free, land conserving, and locally controllable.

Costs, as we have seen, inhibit widespread use of PVs. The example in the previous section was for a household with minimal needs (960 W) at the most. A more typical household may have peak loads of 30,000 W. The price of the system for that household would be at least $10,000, which is appreciable compared to the price of the home. The corresponding annual cost of a loan would be more than what an electric utility charges. On the other hand, prices of PV devices are decreasing as experience is gained in their manufacture and as ways are found to generate more current from a single cell [1]. Also, experience will also result in lower system costs. The advantages of PV devices seem so strong that their eventual success appears inevitable; however, their comparative cost advantage depends on the price of oil. If oil prices remain low, research on PVs will be of low priority, so they will not develop quickly.

In countries without an extensive power grid, PV power generation makes much sense. Water pumps in remote regions, electric fences to keep elephants in national parks, and power for village centers, in addition to hospitals and communication gear, are applications in which PVs are being used. Alternative ways of generating electricity in such applications are diesel generators, which are noisy, smelly, and use imported fuel that is difficult to transport, or wind and hydro, which are not always available. Solar electric systems will work unattended for long periods of time — the biggest problem is usually the battery. A small amount of electric power can often make a significant difference in a remote application and PVs can produce this small amount of power. The cost of alternatives is usually much higher than that of PVs in those applications. Photovoltaics, of course, have had extensive use in the most remote application of all — spacecraft. Until the price decreases, solar cells may not be cost efficient in many other applications; however, their potential is clearly evident.

## REFERENCES

1. Carts-Powell, Y. (1997, December). Solar energy closes in on cost-effective electricity. *Laser Focus World*, 67–75.

   *Encouraging update on costs, includes list of helpful web sites.*

2. IEEE, *Spectrum*. IEEE, 345 East 47th Street, New York, NY 10017.

   *This journal has included articles on photovoltaics which can be understood by people with some science background. The February 1980 issue includes a set of articles which cover both physics and systems. The September 1981 issue has an encouraging article on applications.*

3. Loferski, J. J., Ahmad, J. M., and Pandey, A. (1988). Performance of photovoltaic cells incorporated into unique hybrid/photovoltaic/thermal panels of a 2.8-kW residential solar energy conversion system. 1988 Annual Meeting, American Solar Energy, Inc., June 20–24, Cambridge, MA.

*Describes house with integrated heating and electric power generation.*

4. Maycock, P. D., and Stirewalt, E. (1984). *Photovoltaics — Sunlight to Electricity in One Step*. Brick House, Andover, MA.

   *Covers technology, applications, and economics in a readable way.*

5. National Center for Appropriate Technology (1984). *Homemade Electricity*. For sale by the Superintendent of Documents, U.S. Government Printing Office, Washington, DC 20402.

   *The photovoltaic section is 6 pages long and is chiefly a description of various projects, including the farmer's water system at the beginning of this chapter.*

## PROBLEMS AND PROJECTS

1. Photovoltaic devices make it possible to have electric energy in remote places. Propose a possible application not currently in use, for example, power for hikers in mountains. How much cell area would be required? (Start by estimating needs.) Sketch the system.

2. Sketch the layout, that is, the cell connections of a solar panel giving 120 V DC and 10 A.

3. The most common failure for a solar cell is for it to become an open circuit; that is, it does not produce any current nor let any current flow through it. Show how the connection in Fig. 7.4a is more reliable than that in Fig. 7.4b.

4. How much battery capacity (in ampere-hours) and how big a cell array would you need to get power for 5 cloudy days in a row if you used 4800 W-hr a day?

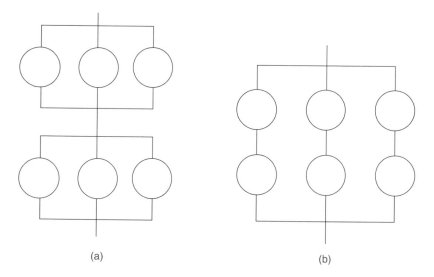

(a)                                                    (b)

**Fig. 7.4. Alternative connections of cells.**

5. Estimate the cost of a photovoltaic system producing enough power for a home using 5000 W for 18 hr a day. If you want to use the power company for storage, assume it buys electric power for half the price it sells it. If you want to use batteries for storage, then use costs in Chapter 3.

6. Proponents of photovoltaic systems state that the roofs of the United States have sufficient area to hold the number of solar cells required to supply all residential use. If that use is $2.4 \times 10^8$ kW do you agree?

7. Proponents of photovoltaic systems point to the additional advantage of not having every one connected to the same centralized system. Argue in favor of and against this advantage.

# APPENDIX

## *Introduction to Physics of Semiconductors*[1]

### *The Flow of Currents in Metals*

The reader undoubtedly knows that all substances are composed of atoms and that atoms in turn are composed of electrons, protons, and neutrons. The electrons can be thought of as being small particles with a negative charge that surround the nucleus. The nucleus is composed of protons, which have a positive charge, and neutrons, which have no charge. There are just as many protons as electrons in an atom and because an atom has just as many electrons as protons it is electrically neutral. The protons and neutrons composing the nucleus are much heavier than the electrons and do not move.

The structure of a piece of metal is indicated in Fig. 7.5. The circles represent the nuclei. The dashed lines represent electrons. The atoms making up metals have one or two electrons which are located relatively far from the nucleus. These electrons are called "valence" electrons. The electrons which are not valence electrons are closely associated with the nucleus and neutralize some of the protons of the nucleus. The valence electrons are not closely associated with the nucleus and can drift away from it. It is convenient, therefore, to consider a piece of metal as a collection of nuclei, each with a positive charge, with electrons floating in the space between the nuclei. The floating electrons are valence electrons which have drifted away from the nuclei and are therefore not associated with any particular atom. When an electron drifts away from an atom, the nucleus of that atom is left with an unneutralized proton; thus, that nucleus has a net positive charge. Of course, the piece of metal as a whole has no

---

[1] This section is based on B. Hazeltine, *Electronic Circuits and Applications*, Chapter 2, Kendall Hunt, Dubuque, IA, 1978.

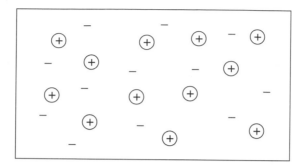

**Fig. 7.5. Atomic structure of a piece of metal.**

net charge because for every unneutralized proton in a nucleus, there is an electron floating somewhere in the metal.

Using this view of the interior of a piece of metal, it is easy to see how electric current is conducted. If we attach a conductor between the terminals of a battery, as in Fig. 7.6, we see that electrons near the positive battery terminal will be attracted to the terminal. This means that the positive nuclei in the region where the electrons were will not be neutralized.

Thus, the region near the positive battery terminal will become slightly positive. Electrons from other parts of the conductor, further away from the positive battery terminal, will be attracted to this positive region. The region these electrons leave will become positive, which will permit other

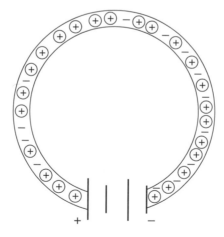

**Fig. 7.6. A conductor connected to a battery.**

electrons still further away to flow toward the positive terminal of the battery. The result is that there is a continual flow of electrons through the entire conductor away from the negative battery terminal and toward the positive terminal. At the positive terminal these electrons enter the battery, which pushes them out of its negative terminal. According to our model the battery energy is spent pushing the electrons out. (The reader should recall that according to the usual definition, current flows out of the positive terminal of the battery into the negative terminal. The usual definition of current was created before people realized current was carried by negative charges.)

### Solar Cells

Solar cells are usually made of crystalline germanium or crystalline silicon. (In our discussion, we will consider silicon, but the results apply equally well to germanium.) Because the material is crystalline, the nuclei of the atoms are arranged in a regular pattern (Fig. 7.7).

In silicon (as well as in germanium) each nucleus has space for eight valence electrons, but each atom has only four valence electrons. Thus, by sharing valence electrons between adjacent nuclei, each nucleus can fill all its spaces. Normally, at room temperature nearly all the valence electrons are shared between neighboring nuclei and tend to stay at the same location.

There are a few valence electrons, however, that have enough energy to get away from these places of sharing, and these energetic electrons float through the interior of the semiconductor just as the valence electrons do in a metal. Because there are so few electrons which are mobile at room temperatures, only a small amount of current will flow through the semiconductor at room temperatures. (This is the reason the name "semiconductor" is used.) If the material is heated up, however, many of

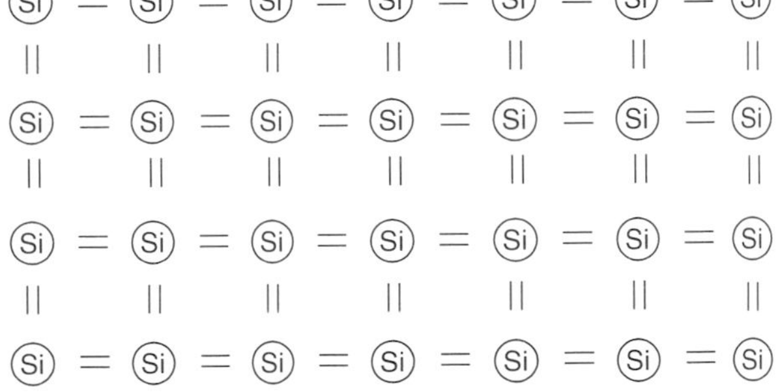

**Fig. 7.7. Internal structure of semiconductor material.**

the valence electrons obtain enough energy to escape from their locations and, therefore, the resistance of the material decreases markedly at high temperatures.

### p-Type and n-Type Impurities

Adding impurities to the silicon allows appreciable current to flow at room temperature. As was mentioned, the reason there are only a few electrons available at room temperature to carry current is that each silicon atom has four valence electrons and has space for eight valence electrons. By sharing valence electrons each atom fills its eight spaces. Now we assume a small amount of arsenic is mixed with the silicon when the silicon is molten. The arsenic atom is basically similar to the silicon atom and, therefore, fits right into the regular crystal structure. The arsenic atom differs from the silicon atom in one respect, however; it has five valence electrons. Of course, it can only share four of these valence electrons with neighboring silicon atoms; the remaining electron is free to wander through the crystal. An impurity such as arsenic which adds free electrons to the material is called an "n-type" impurity — the "n" refers to the negative charge of the electron.

There are also "p-type" impurities which add positive "holes" to the material. A p-type impurity such as gallium has only three valence electrons, one less than silicon. Therefore, when gallium is mixed with silicon, the gallium atoms will not have enough electrons to share with the neighboring silicon atoms. This is illustrated in Fig. 7.8, in which a gallium atom is shown surrounded by silicon atoms. (The chemical abbreviation for silicon is Si and for gallium is Ga.) Figure 7.8 demonstrates that an electron is missing from the otherwise periodic structure of the crystal. The place where the missing electron should be is called a hole. (The hole in Fig. 7.8 is directly above the gallium atom.) In a material containing p-type impurities, it is possible for an electron associated with another silicon

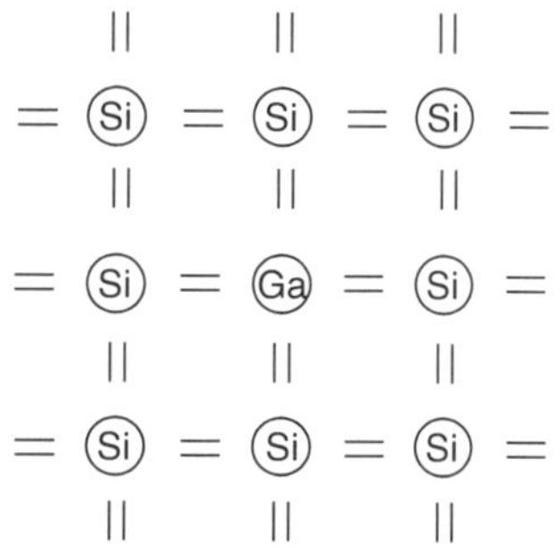

**Fig. 7.8. Silicon with gallium impurity added.**

atom to slip into a hole, thus filling it. If this happens, another hole is left in the space from which the electron came. Not only is a hole produced in the crystal structure but also one of the positive charges on the nucleus of an atom adjacent to this new hole is no longer neutralized. It is convenient to consider this positive charge as being associated with the hole itself. When an electron from another atom slips into this new hole, it leaves a net positive charge somewhere else; thus, in effect, the hole migrates through the material, acting like a particle carrying a positive charge.

### Bound Charges

Not only do the moving holes and electrons carry a charge but also the region from which the holes and electrons came have a charge. The nucleus of the gallium atom contains just enough protons to neutralize its own electrons. Therefore, when a hole leaves a gallium nuclei, that is, when an electron slips into the hole adjacent to the gallium nucleus, it adds a negative charge to the region near the gallium atom. The negative charge is associated with the nucleus and it is not free to move through the conductor. It is therefore called a "bound" charge. The entire crystal is still electrically neutral because, as we have seen, the hole produced when the electron slipped into the hole near the gallium nucleus is positive.

In a material containing an n-type impurity, the impurity adds an extra electron to the crystal; when this extra electron leaves the neighborhood of the arsenic atom from which it came, (this region takes on a positive charge because one of the protons in the arsenic nucleus is no longer neutralized by a valence electron. These charges, bound in a region by the presence of impurities, are the basis of the operation of semiconductor diodes and solar cells.

### The p–n Junction

Silicon with an n-type impurity added is called n-type silicon. Similarly, silicon with a p-type impurity is called p-type silicon. Because holes predominate in p-type material, they are called "majority carriers." Electrons are "minority carriers" in p-type materials. In n-type material the designations are reversed. Electrons predominate so they are the majority carriers. There are only a few holes in p-type material so they are minority carriers.

We now discuss what happens when a piece of p-type silicon is joined to a piece of n-type silicon forming a so-called "p–n junction," as illustrated in Fig. 7.9a. The p-type material is shown as being full of positive holes, whereas the n-type material contains negative electrons. The holes in the p-type material and the electrons in the n-type material are, of course, moving. When the p-type and the n-type materials are joined, therefore, some holes will go from the p-type silicon to the n-type and some electrons from the n-type to the p-type. An electron, when it enters to the p-type material, will probably encounter a hole fairly soon because there are many holes in the p-type material. It is apparent that when an electron encounters a hole it will slip into the hole. When this happens, both the hole and the electron disappear in the sense that neither is available to carry current. The result is that, shortly after the two materials are joined, in the p-type material there are fewer holes near the junction than elsewhere. Similarly, in the n-type material, there are fewer electrons near the junction than there are elsewhere in the material because the electrons near the junction have combined with holes from the p-type material and vanished. The region near the junction where there are few majority carriers is called the "depletion region" (Fig. 7.9b).

When the electrons in the n-type material near the junction disappear by combining with holes, the depletion region in the n-type material takes on a positive charge. The origin of this positive charge is the fifth proton in the nuclei of the impurity atoms — the bound charge mentioned previously.

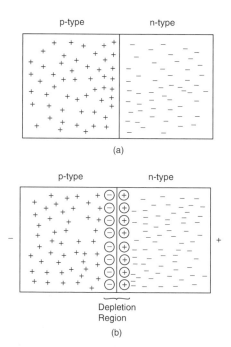

**Fig. 7.9. (a) p–n Junction when first formed. (b) p–n Junction when steady state is reached.**

Originally, these protons were neutralized by electrons wandering through the crystal; when the free electrons disappear, the region takes on a positive charge. The bound positive charges are circled in Fig. 7.9b.

The result of combining a piece of n-type material with a piece of p-type material is that the p-type material near the junction becomes negatively charged and the n-type material near the junction becomes positively charged. In other words, a voltage is produced across the junction with the n-type material at the higher voltage. The polarity of the p-type material and of the n-type is indicated in Fig. 7.9b. An easy way to remember which material becomes positive and which becomes negative is to recall that the n-type loses electrons to the p-type and therefore becomes positive. The p-type gains electrons and therefore becomes negative.

Such a voltage is produced whenever two semiconductors are joined. In fact, even when two different metals are joined, a voltage is produced because the electrons in one metal will be more energetic than the electrons in the other metal and, therefore, there will be a net flow of electrons from one piece of metal to the other. The piece of metal which acquires more electrons than it loses will be at the lower voltage. This potential difference produced when two dissimilar metals are joined is called the "contact potential."

The voltage which is established between the n-type and the p-type materials prevents further flow of electrons and holes because the n-type material, having become negative, repels the further flow of electrons. Thus, a very short time after the junction has been formed, no more holes or electrons will flow across the junction.

### Photovoltaic Devices

A PV device is a p–n junction (Fig. 7.10). Electrical power is created because energy from the sun, or any light source, gives electrons sufficient energy to leave their bonds. The free electrons in the p-type material, which are minority carriers, move to the n-type material and then to the load. The n-type material collects the electrons because it is positive.

Commercial solar cells can be made in different ways. A typical way is to deposit a layer of silicon on a metal backing. Impurities are introduced in two stages so a p layer is produced adjacent to the backing and an n layer on top. Energy from the sun passes through the n layer and liberates electrons within the p layer. These electrons move to the n layer, where they are collected by thin metal strips on the surface. The external load is connected to these strips and the backing.

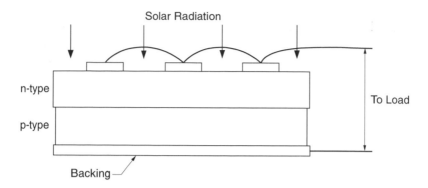

**Fig. 7.10. Photovoltaic device.**

*Semiconductor Diode*

A diode allows current to flow in only one direction. It is basically a p–n junction. Assume a battery is connected across the junction with the positive terminal of the battery connected to the p-type material and the negative terminal connected to the n-type region, as in Fig. 7.11. When the battery is connected, the direction of its voltage will counteract the voltage developed across the junction and thereby allow electrons from the n-type region to move into the p-type region. A small voltage — just sufficient to overcome the voltage at the junction produced by the layers of bound charge — will allow a fairly large current to flow across the junction because there are many electrons in the n-type material.

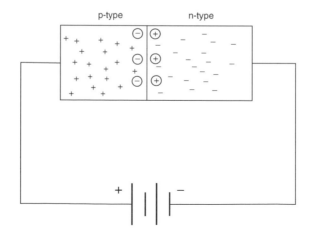

**Fig. 7.11. p–n junction and battery — current flowing.**

If the polarity of the battery shown in Fig. 7.11 is reversed, that is, if the positive terminal is connected to the n-type region and the negative terminal to the p-type region, then the battery voltage will add to the voltage produced across the junction (Fig. 7.12). (Recall that the p-type material became negative and the n-type became positive when the junction was formed.)

Because the battery voltage adds to the voltage established by the initial flow of holes and electrons, it will be even harder for holes to move into the n-type material and for electrons to move into the p-type material. Therefore, only a very small current will flow. (The current that does flow is produced mostly by minority carriers, holes in the n-type region, and electrons in the p-type region. These minority carriers are the result of electrons which have found enough energy, perhaps from heat, to escape from the valence bounds.)

The implication of our discussion is that the current flows only when the applied voltage makes the p-type material function at a more positive voltage so electrons can move into the p-type area and holes into the n-type area. When the applied voltage makes the n-type material positive, current does not flow because electrons "want" to be at the more positive voltage and holes are repelled by positive voltages. This device is called a diode. It allows current to flow in only one direction.

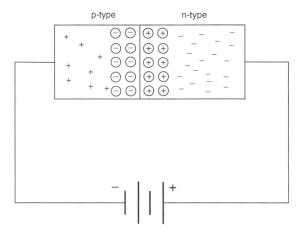

**Fig. 7.12. p–n junction and battery connected—no current flow.**

# *8*

·········

# METHANE DIGESTERS

··················

**Sign reads: "Biogas Generation Project Experiment 3:
Structural and feed processing innovations for household size digestors to
improve ease of loading and for stable, longer gas production period."**

A methane digester is basically a large tank in which manure and other plant wastes ferment, producing biogas — a fuel that burns — and an odorless effluent, which is usable as fertilizer. Biogas can replace natural gas or gasoline and the effluent is easier to apply as fertilizer than is raw manure. The fermentation process is not difficult to operate.

An early proponent was a pig farmer, John Fry, who realized that he was spending most of his time shoveling manure from the barn to a compost heap and from the compost heap to the field. He built a digester primarily to reduce shoveling but found that the amount of gas generated was sufficient to power all the machinery used on the farm. We study methane digestion because it is a simple technology which produces useful by-products — fuel and fertilizer — and improves the environment by processing wastes. The following sections describe how the process works and how a digester is built.

# THE SCIENCE FAIR CONTEST

In October 1980, Javed Ahmed, a senior at St. Francis Xavier's High School in Dhaka, Bangladesh, decided that he would enter an interschool science fair. There was considerable rivalry among the schools participating in the science fair and for the previous 2 years St. Francis Xavier's had won first prize. Javed had to develop a project never presented before at a high school level. He also wanted his work to be more than just another science fair project. He specifically wanted to do something innovative concerning the energy crisis in Bangladesh, one of the poorest countries in the world. Its natural resources are natural gas, tea, jute (a fiber), and people (population 100 million, of which 80% live in small villages). Bangladesh is located between India and Burma and is about the size of Wisconsin. Much of the land is the delta of the Ganges and other rivers. The science fair was going to be held in 6 weeks, and an idea had to be formulated quickly.

Two days later, Javed read an article describing the success of biogas production in India. Biogas is the gas that is obtained when waste products, such as livestock manure, are fermented. This gas contains methane and thus can be used as a fuel gas. Javed thought it may be possible to produce biogas on a small scale so that individual farmers could construct a small biogas generator providing gas for individual households or farming purposes. After much research at the library and several visits with research workers at Dhaka University, Javed gathered enough information to design a small biogas generator. He kept his design simple so that it could be built with a farmer's limited funds. As excited as he was, there was one problem — Javed was totally broke. Therefore, he spoke to his school principal.

The principal agreed to give school funds for the project but under one condition: that the biogas generator be powerful enough to supply the school's science laboratory with sufficient gas for its Bunsen burners to be used during class — 3 times a week for 45 min each. St. Francis Xavier's had a small laboratory but had not yet acquired the gas connection for the Bunsen burners; an order for a gas line had been placed but installation was not expected for 3 months. If everything went according to plan, the biogas generator could supply the lab for the 2 months after the science fair. Now, not only did Javed's group (three of his fellow classmates had joined him in the project) have a time constraint but also he absolutely had to make the project work. He had 4.5 weeks to build a working model of a biogas generator before the science fair. Even if biogas is needed in a school, the reader is probably asking whether Javed's project would have any relevance for a farmer. What use would a farmer have for biogas?

## A FARM IN BANGLADESH

We examine the situation of a farmer who might use biogas. The farmer would typically have three or four children, one or two of whom may be old enough to work. A 12-year-old boy can work full-time in the fields but earns only about one-third of what his father earns. The family eats twice a day—at noon and at dusk. The staple food is rice. Earthen clay wood-burning stoves are used for cooking. For light at night, kerosene-burning lamps or lanterns are used because little electricity is supplied to the rural areas of Bangladesh.

The farmer would use biogas mainly for cooking and lighting. If the biogas generator were big enough, the gas could be used to drive a water pump to provide running water. A big digester could also fuel an electric generator which could run electric fans, moderating the hot and humid climate. A small generator would produce enough fuel for cooking and lighting for two rooms. Before we discuss how Javed and his friends went about constructing the generator, we discuss the basic theory behind the production of biogas.

## WHAT IS BIOGAS?

Biogas is primarily methane (hence, biogas generators are also called methane digesters), with some carbon dioxide, carbon monoxide, and traces of other gases. Methane consists of carbon and hydrogen. The chemical symbol is $CH_4$. The generation process employs bacteria that work under anaerobic conditions, that is, without oxygen—oxygen kills the bacteria. The raw material commonly used is livestock manure, straw, weeds, sewage wastes, and so forth. The methane-producing bacteria are found naturally in all these raw materials, especially in livestock manure.

There are several reasons why biogas technology is useful as an alternate form of technology: Primarily, the raw material used is very cheap and to farmers it is practically free; the biogas can be utilized for many household and farming applications and the effluent obtained from the digester is a good fertilizer for crops because it has not lost any of the nutritional value of the original raw material but is odorless and usually germ free; the burning of biogas does not produce harmful gases so it is environmentally clean; and the technology is relatively simple and can be reproduced in large or small scale, by many people, without the need for a large initial capital investment [5].

## *Raw Material*

There are several conditions in the tank that must be maintained for efficient gas production. The bacteria need nutrients. These are provided by the raw material, which can be a combination of things, for example, cow manure, straw, and night soil (human wastes) combined with water. Essentially any organic material can be used as input except for mineral oil and lignin (the "woody" part of plants and trees), neither of which will ferment. Animal manure, sludge, and effluents from fermentation factories, such as leather processors, are among the organic materials that easily produce biogas. Some materials, such as leaves, stalks, and straw, which are high in cellulose content, are harder for the methane-producing bacteria to "digest." Such material can be pretreated, however, and then fed in as raw material.

There are two ways to pretreat raw material such as straw and leaves: (i) feeding it to animals and using the resulting manure for the biogas generator or (ii) composting the material by mixing it with manure and a small amount of limewater (to make it less acidic) and leaving the mixture in a heap for several days. Either process will break up the cellulose in the stalks. Without this pretreatment, the digestion in the generator will be slow and the production of biogas will decrease considerably. It is evident that the first method of pretreatment is easy and has other benefits, such as better fed animals.

The raw material is the nutrient for the bacteria and it is important that the bacteria's diet be balanced. The raw material must contain carbon and nitrogen. The methane-forming bacteria use the carbon for fuel and the nitrogen for building their cell structures, so the carbon to nitrogen ratio ($C:N$) is important. The bacteria need about 25–30 times more carbon than nitrogen. In other words, the $C:N$ ratio of the raw material should be approximately 25–30 : 1. Table 8.1 provides a list of raw materials and their $C:N$ ratios. The combination mentioned earlier has a $C:N$ ratio approximately 27 : 1 [cornstalks; 53 : 1; cattle manure; 25 : 1; human feces; 2.9 : 1 — a mixture of equal parts by weight will give an average $(53 + 25 + 2.9) : 3 = 27 : 1$]. To summarize, optimum biogas production requires proper raw materials in the proper ratio. The raw materials

### Table 8.1. C:N Ratio of Common Raw Materials

| Raw | C (dry wt %) | N (dry wt %) | C:N ratio |
|---|---|---|---|
| Stalks | | | |
| Straw (wheat) | 46 | 0.53 | 87:1 |
| Straw (rice) | 42 | 0.63 | 67:1 |
| Stalks (corn) | 40 | 0.75 | 53:1 |
| Fallen leaves | 41 | 1.00 | 47:1 |
| Stalks (soybean) | 41 | 1.30 | 32:1 |
| Weeds | 14 | 0.54 | 27:1 |
| Stalks and leaves (peanut) | 11 | 0.59 | 19:1 |
| Manure | | | |
| Sheep | 16 | 0.55 | 29:1 |
| Cattle | 7.3 | 0.29 | 25:1 |
| Horse | 10 | 0.42 | 25:1 |
| Swine | 7.8 | 0.65 | 13:1 |
| Human feces | 2.5 | 0.85 | 2.9:1 |

should also be diluted with water approximately 50:50 so the mixture is fluid.

## Managing the Process

There is a close relationship between biogas production and temperature because the efficiency of methane-producing bacteria work depends on the temperature. There are two types of methane-producing bacteria: thermophyllic and mesophyllic. The first type of bacteria thrive at temperatures between 47 and 55°C, whereas the latter type of bacteria operate best between 27 and 38°C. Thermophyllic bacteria produce gas at a daily rate of 2.5 m$^3$ gas/m$^3$ digester volume at their optimum temperature, whereas mesophyllic bacteria produce 1.0–1.5 m$^3$/m$^3$ per day at their optimum temperature. Although thermophyllic bacteria produce biogas at a higher rate, it may be costly to keep the generator heated. Some heat, but not enough, is produced by the digestion. Thus, mesophyllic bacteria are used in most digesters.

Even mesophyllic bacteria produce faster when the temperature is above normal outdoor temperature. At 25°C a typical generator will produce about 1 m$^3$ of gas per day for each cubic meter of tank volume.

At 15°C, the production rate decreases to about 0.3 m$^3$ gas per day. Cattle manure is completely digested in a generator at 24°C in 50 days. At 35°C, the fermentation period is only 28 days. Thus, by raising the temperature by 11°C, one can almost double the rate of gas production, but the total amount of gas is about the same. A faster rate is useful, even if more gas is not produced, because a faster rate means a smaller volume tank is needed for the same output. A digester can be heated by burning some of the gas produced, but this partly defeats the purpose if the gas could be used elsewhere. Solar heating is possible. To take advantage of the sun's heat to the fullest extend, the digester should be kept in a sunny place, with a wind barrier around it. Insulation can be used, which keeps heat produced inside the digester but also keeps the sun's energy out.

Another factor influencing the operation of the digester is pH, a measure of acidity: Lower pH corresponds to higher acidity. Biogas fermentation takes place in a slightly alkaline environment (pH value between 7.0 and 8.0). It has been observed that during the initial stage of biogas fermentation, the pH level drops slightly before it stabilizes, that is, the mixture becomes acidic. During this initial stage, it may be helpful to add alkaline materials to increase the pH level to between 7.0 and 8.0. Adding burnt ashes to manure is one way of raising the alkalinity of the digestive material. Limewater may be used, but care must be taken to use a very dilute solution; otherwise, the pH level increases to more than 8.0 and methane-producing bacteria are killed. Generally, the pH level adjusts to within the functioning range after the initial stage. If a digester is not producing as much gas as expected, then the pH may be too low and ashes should be added.

Each time new material is fed to the digester, the rate of biogas production drops slightly. The reason for the decrease in gas production is that the raw material entering the digester contains relatively few methane-producing bacteria, whereas the sludge that is removed from the digester is rich in such bacteria. To compensate for the reduction in bacteria after feeding, one may prerot the raw material by piling it in a heap mixed with a small amount of sludge from a digester. This process of prerotting the raw material with "starter" bacteria is called enrichment.

Adding bacteria with new raw material makes a big difference. A large-scale biogas generator will not produce any methane until 4 weeks after it is filled with fresh raw material. On the other hand, when prerotted raw material is used along with some digester sludge, the methane content of the gas generated will be 50% of full production on the sixth day and 75% by the end of 4 weeks. Besides sludge from other digesters, sewage from slaughterhouses (or any house for that matter) is a common additive to introduce the bacteria. Using additives greatly reduces the retention period and increases the early gas yield and its methane content.

It is not hard to visualize the inside of a biogas generator. The contents divide themselves into three layers: The raw material settles in the bottom of the digester, a murky watery layer in the middle contains the bacteria, and the third (top) layer is a thick crust of scum. This layer prevents the formation of gas at an optimum rate because when the gas is formed, it has trouble escaping to the gas collector through the crust. The remedy is stirring the digester frequently without, of course, letting in oxygen. Stirring ensures an even distribution of raw materials, extends the contact surface area of the raw materials with the bacteria, and speeds up fermentation, thereby increasing the gas yield. Experiments show that stirring can increase the gas yield by 10–15%.

The pressure of the gas inside the digester is important. Anyone who has opened a bottle of carbonated drink knows that once the cap is removed, the drink becomes visibly "fizzy" as pressure is released inside the bottle. Before the cap was opened, the pressure inside the bottle prevented the dissolved carbon dioxide from forming bubbles. Pressure can play a similar role inside a biogas generator: High internal pressure will not let gas bubbles form. Internal pressure $<5\%$ more than atmospheric pressure will not affect biogas production significantly; however, any increase in pressure beyond this will reduce gas production. Thus, it may be useful to use a manometer (a pressure measuring device), which is attached to a biogas generator. A way to control the internal pressure should also be included in the design of a generator. One way of doing this will be discussed later.

## Problems in Digester Operation

One may design and build a biogas generator and feed it raw material properly but still obtain no methane in the gas. If, after checking the generator for leaks and the raw material for correct feeding, one still finds no production of gas, then the digester may have been contaminated with fermentation inhibitors, perhaps from something sprayed on the field from which the straw was taken. Biogas formation is a microbiological process and certain materials may inhibit the biological activity of the microbes. If these "poisons" contaminate the digester badly enough the tank must be emptied and washed. Some inhibitors, along with their inhibiting concentrations, are shown in Table 8.2. Care should be taken regarding where and what sort of raw material is fed into the digester. None of these inhibitors are likely to be in manure or household wastes.

Another type of "poisoning" may occur if the concentration of volatile acids released by the fermentation is too high. The concentration of volatile acids may rise for several reasons: (i) too high a concentration of the raw

### Table 8.2. Biogas Inhibitors

| Substance | Inhibiting concentration (mg/liter) |
|---|---|
| Soluble sulfides | 200 |
| Sodium cation | 8,000 |
| Potassium cation | 12,000 |
| Calcium cation | 8,000 |
| Magnesium cation | 3,000 |

materials and not enough diluting water; (ii) lack of methane bacteria; (iii) too much acidic raw material, e.g., manure with a high proportion of cow urine; or (iv) too little fresh raw material being fed into the digester, making the mixture inside stagnate.

A major possible problem is an explosion. Care must be taken to keep air away from the biogas until one wants to burn it. Leaks must be avoided. A big explosion could be fatal and a small one certainly would leave a disagreeable mess.

Now that some basic background on biogas production has been given, we discuss how Javed's group actually carried out their project work.

## JAVED'S DESIGN

First, Javed had to make the digester in such a way that he would have a fermentation chamber and a gas collection chamber separated from each other (Fig. 8.1) [3]. For the two chambers, he used two 55-gallon drums and a smaller 20-gallon drum. The fermentation chamber was formed from one of the 55-gallon drums, and the gas collection chamber was made from the other two drums. He collected the gas over water. Gas collection over water is easy and has other advantages. Not only can one regulate the internal pressure of the fermentation chamber by adjusting the weight of the smaller drum but also any other gases in the biogas, especially carbon dioxide, are dissolved in passing through water. Methane is only slightly soluble so water is an effective filter.

The fermentation chamber has an input pipe for the input of raw material and an output pipe for the removal of used sludge. Used sludge is dense and settles at the bottom of the fermentation chamber. The outlet pipe thus comes out from the lower portion of the chamber. The input pipe was placed near the top of the chamber but below the fluid level to prevent the escape of gas.

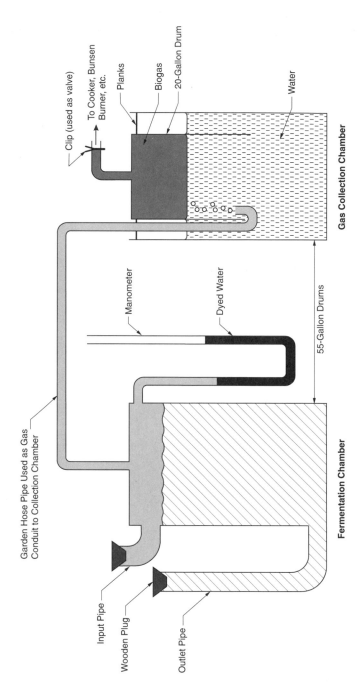

**Fig. 8.1. Javed's methane digester.**

Both the input and output pipes were made in two pieces. First, steel cylinders 6 in. long and 4 in. in diameter was welded to the 55-gallon drum. (Welding is an often used craft in rural Bangladesh.) To these pipes, thick flexible tubing were snugly attached. Flexible tubing was used so that when it is necessary to close the inlet or the outlet, the tubing is simply lifted up and the free end plugged with a wooden plug. When sludge is to be let out, the outlet pipe is unplugged and the flexible tube lowered. Pressure from the weight of the liquid and from the gas pushes the sludge out when the pipe is lowered. When fresh raw material is poured in, the input pipe is kept at an elevated angle to avoid gas escape. After the raw material is poured in, the inside of digester can be stirred using a thick, long bent wire, which also breaks up the crust. The inlet pipe is thus used for the input of raw material and for frequent stirring of the mixture.

## CONSTRUCTION OF THE DIGESTER

The first step in construction was painting. Since the inside of the digester was prone to corrosion, Javed's group decided that it would be best if they painted the inside of the fermentation chamber with red lead oxide paint. To do the painting the top cover of the drum was cut off using a steel cutting saw. The inside was painted with the lead oxide and the top was welded back on. Before the top of the 55-gallon drum was welded back on, the openings for the inlet and outlet pipes were made. A 0.5-in.-diameter hole was made with a cutting blowtorch and a 4-in.-long pipe, 0.5 in. in diameter, was welded on. This pipe would be the outlet for the gas.

The digester tank would need a manometer. At a level slightly higher than the input pipe a 0.5-in. hole was made, and a 2-in. metal pipe, 0.5 in. in diameter, was welded on. A glass U-tube was attached to this pipe. The U-tube was to act as a manometer. The smaller arm of the U-tube, which was attached to the pipe, was about 16 in. long, and the other end was more than 32 in. long. Thus, the total length of the glass tube was about 60 in. long, allowing for the curved bottom part of the tube and the place where the tube was attached to the pipe. The tube was filled with dyed water about 16 in. high on each arm. At an internal pressure of 1 atm the water levels in the two arms would be the same. When the internal pressure increased, the water level in the outer arm would rise and that in the closer arm would decrease until the pressure created by the column of water in the outer arm was equal to the extra pressure inside the fermentation chamber. From calculations made using the density of water it was found that an internal pressure 10% higher than atmospheric pressure would produce a water column 40 in. high in the longer arm. Such a high internal pressure would have to be reduced by letting gas flow into the collection chamber.

The gas collection chamber, as mentioned earlier, would collect over water. In the construction of the collection chamber, the top of the second 55-gallon drum was cut and discarded. Since the drum was to be filled with water, it was prone to corrosion as well, especially since some of the gases to be dissolved in the water were acidic. Thus, the inside of this drum was also painted with red lead oxide paint. The 20-gallon drum was also painted with the lead oxide on both the outside and the inside. Before it was painted, though, the top of the 20-gallon drum was removed and the whole drum was inverted. A 0.5-in. hole was made at the bottom of the drum (which would now be used as the top of the collector drum). Again a 4-in. long, 0.5-in.-diameter pipe was welded to the hole. This was to be the outlet of the gas from the collector, leading to wherever the gas was to be used. A flexible rubber pipe was attached to the cylinder and, as a valve, a strong clip was attached to the rubber pipe. At the top of the gas collector, a small metal strip was attached on the inside (Fig. 8.2). This was be the clamp retaining the pipe leading from the fermentation chamber to the collector chamber. The hose used to convey the gas from the fermentation chamber to the collector was a regular garden hose — cheap and flexible. Care had to be taken to ensure that the hose did not kink and prevent a good flow of gas.

Collection over water requires that the collector chamber initially be filled with water. This was easily done by opening the valve and lowering the collector chamber into the water until all or most of the air had been expelled by the water. Once the collector chamber was full of water, the gas outlet valve was closed and a check was made that the valve did not leak so that the collector chamber would not be contaminated with air. To prevent its sinking, the collector chamber was supported by two wires passing under the collector attached to two wooden planks on top of the water drum (Fig. 8.3). These wires not only provided stability but also controlled

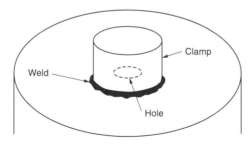

**Gas Collection Chamber**

**Fig. 8.2. Gas outlet.**

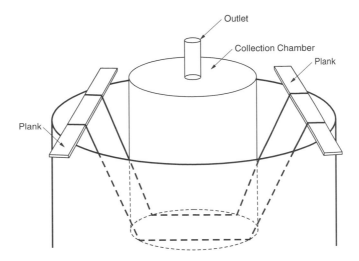

**Fig. 8.3. Pressure control for gas collection chamber.**

the internal pressure of the fermentation chamber. If the internal pressure were too high, then the wooden planks could be turned over, decreasing the length of the wire. Shortening of the wire would raise the drum and this would cause the internal pressure to drop. Conversely, if the pressure of the collected gas needed to be increased when the gas was being used (e.g., to have a more powerful flow of the gas to the burners), then the length of the wire would be increased by turning the planks in the other direction, and bricks would be placed on top of the collector to push it down into the water.

## TESTING THE DIGESTER

Once the generator was set up, it was ready to be tested. The first step was to collect raw material for fermentation. A proper combination of raw materials needed to be made. After collecting manure, water was added to it. Since the manure was not dry, the ratio of water to manure was set at $1:1$. If the manure had been dryer than fresh manure, a ratio of $3:2$ (water to manure) would have been used. Javed's group got the ash to make the pH of the raw material slightly alkaline from the owner of a house across the street from the school who still used wood-burning clay stoves. They did not know exactly how much ash to use; as a reasonable estimate, they put in a handful of ash for every 5 gallons of raw material.

Now the group was ready to start the digestive process. For smooth and safe operation, it was necessary to make the whole system airtight. Thus, the fermentation chamber was filled entirely with water. If any leaks occurred, they could be spotted. After leak testing, the raw material was fed to the digester. The manometer was filled with water dyed red to make it easier to read. The outside of the fermentation chamber was painted black to absorb the sun's energy and the location of the digester was chosen so that during the day it would get as much sun as possible. At night, they covered up the fermentation chamber with jute sacks to keep in the warmth gained during the day.

The construction of the entire digester took 3 weeks; during which time they shopped for the material, school midterm exams occurred, and there was some lack of communication among the group (e.g., showing up hours late; or not showing up at all and not letting the others know that one had other pressing engagements). Finally, the digester was filled with the raw material, and the gas collection chamber was ready to trap any evolving gas. The manometer read 1 atm of pressure inside the fermentation chamber. The science fair was 1 week away. Javed's group had to wait patiently until enough methane formed so that the gas would burn.

During the first 2 days there was no gas production. On the third day, however, some gas was obtained, but it did not burn. Javed assumed that the methane was diluted too much by the air that was already present in the fermentation chamber. On the fourth day, the gas lit up and the four students started celebrating! During the next 3 days, 10 gallons of gas were produced. When the digester was exhibited at the science fair, it was half full of combustible biogas.

For their presentation at the fair, the team had to show specifics of the cost to build the digester and that the cost would be feasible for a farmer owning a single cow. The figures they used are presented in Table 8.3. The unit of currency in Bangladesh is Taka (Tk).

This biogas generator can supply gas at the rate of 1000 liters of gas a day per 1000 liters of digester raw material in the tank during moderate weather. During cold weather, the output is decreased by about 30%. A 55-gallon drum can hold approximately 208 liters of raw material, so the daily gas output in moderate weather is about 208 liters. This output was enough to supply the school burners for a week.

## VALUE TO A FARMER

Is this digester of value to a farmer, even a farmer who only wants to use the gas to cook two meals a day? If the farmer uses a small stove for about 60 min a day, then about 280 liters of gas will be needed (see Table 8.6). In

### Table 8.3. Cost of Digester

| Quantity | Item | Cost (Tk)/unit | Cost (Tk) |
|---|---|---|---|
| 2 | 55-gallon drum | 105 | 210 |
| 1 | 20-gallon drum | 55 | 55 |
| 6 ft | Thick rubber piping for input and output | 10/ft | 60 |
| 12 ft | Clear plastic garden hose pipe | 2.50/ft | 30 |
| 1 ft | 0.5-in.-diameter metal pipe | 10/ft | 10 |
| 1 ft | 4-in. diameter metal pipe | 30/ft | 30 |
| 1 Pint | Red lead oxide paint | 20/pint | 20 |
| 2 | Wooden plugs | 2 | 4 |
| | Cost of welding, cutting, and labor charge | 100 | |
| | Transporting material | | 30 |
| | Total | | 549 |

cold weather Javed's digester produces only about 145 liters of gas — enough for only one meal. Therefore, a second fermentation tank, holding more raw material, must be added. The additional cost of this tank is shown in Table 8.4. Another approach to improving gas production is to redesign

### Table 8.4. Costs for Farmer's Digester

| Quantity | Item | Cost (Tk)/unit | Cost (Tk) |
|---|---|---|---|
| 1 | 55-gallon drum | 105 | 105 |
| 6 ft | Thick rubber piping | 10/ft | 60 |
| 12 ft | Clear plastic garden hose pipe | 2.50/ft | 30 |
| 1 ft | 0.5-in.-diameter metal pipe | 10/ft | 10 |
| 1 ft | 4 in.-diameter metal pipe | 30/ft | 30 |
| 2 | Wooden plugs | 2 | 4 |
| | Cost of additional welding, | 1 | |
| | Total additional | | 284 |
| | Cost of original digester | | 549 |
| | Total cost for farmer's digester with two fermentation chambers | | 833 |

the digester, for example, by placing the 55-gallon drum on its side to increase the surface area from which the gas escapes.

The cost is Tk 833. What does this mean for a typical farmer? The per capita national income of Bangladesh in 1980 was $90. The exchange rate of the Taka was Tk 24 = U.S. $1, so an average Bangladesh would earn Tk 2160 per year. Those farmers who could afford a cow and have enough money to eat two meals a day could afford a biogas generator. The average annual income of such a farmer would be approximately Tk 4000. In a village of 500 farmers, about 150 farmers own at least one cow. The others are simply too poor even to think of a biogas generator and concentrate instead on just bringing enough food home for two meals a day.

Thus, a farmer with a yearly income of Tk 4000 has to save a little less than 3 months of income to pay for the digester. It would be possible to save this amount over the period of a year. If the farmer took out a loan from the government-run Agriculture Bank, which allows loans for such purposes, it would be even easier to buy a digester.

The cost can be reduced. It should be noted that some costs in Table 8.4 are very approximate, e.g., transport, and some costs can be reduced considerably if the job is done in bulk, e.g., the cost of the drums and the labor charge. Making digesters may be cheaper if several farmers get together and manufacture them at the same time.

Thus, Javed's group's goal was nearly attained, even though their prototype methane digester would not produce enough for the farmer. They knew their prototype with an extra fermentation chamber would meet the farmer's needs. They did accomplish the goal of operating the gas Bunsen burners in the lab 3 days a week for 45 min. The biogas generator supplied gas for 2 months. Incidentally, Javed's group also was awarded first prize in the science fair competition.

## A Digester Used in China

We briefly discuss a larger scale example of the same technology as that discussed previously [1]. Figure 8.4 shows a cross-sectional diagram of a digester used in the Sichuan Province of China. The internal volume of the fermentation chamber can be between 6 and 123 $m^3$. These sizes are practical for household use. Small plantations and pigsties in communes have digesters with internal volumes between 500 and 1000 $m^3$. The digester in Fig. 8.4 differs from that in Fig. 8.1 in several ways, including the emptying chamber for the effluent.

The digester is built underground using stones, bricks, lime, and cement. It is circular, small, and shallow. There are advantages to placing the digester underground: (i) farmland food can be grown over the tank;

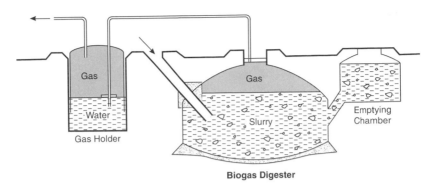

**Biogas Digester**

**Fig. 8.4. Digester used in China. (Reproduced from Renewable Energy Technologies [2].)**

(ii) using the soil as structural support; (iii) improving thermal insulation, thus preventing thermal cracking due to quick changes in temperature and moisture; and (iv) simplifying operation since feeding of the digester and discharging can occur at ground level. The design uses a shallow container to increase surface area for the gas to escape. Of course, the effluent must be pumped out or removed with buckets, although the pressure of the gas raises some effluent into the emptying chamber.

With this design about 1530 liters of biogas is produced each day for every cubic meter of digested material with temperatures between 15 and 30°C. The following are the main components of the digester in Fig. 8.4:

1. A straight inlet, convenient for adding the material without clogging. The rush of material from the inlet also mixes the material inside the digester.

2. An emptying chamber placed so the effluent flows out from the middle of the tank, where 90% of the parasite eggs in the mixture have been killed. The emptying chamber has a volume of about 10% of the digester volume.

3. A removable cover which allows repairs to be made inside a digester. The cover is tested under water when sealing the digester, thus ensuring that the digester is airtight.

In China many villages use methane digesters to decompose night soil. If the digester is well insulated its temperature will exceed 100°F, which is sufficient to kill nearly all the harmful pathogens, so the effluent is generally safe as a fertilizer. One should not use it on leafy vegetables to be eaten raw.

# METHANE DIGESTER FOR A FARM IN THE UNITED STATES

People have experimented with methane digesters for farms in the United States [5]. Assume a farmer has 10 cows and would like to build a digester; how big should it be? How much methane will be produced?

Table 8.5 will be used in our design. To make sure the table is clear, let us consider the top row. It states that one cow will produce 1.5 ft$^3$ of manure each day, that in loading the digester one should add 25% additional water, and that the manure will give 50 ft$^3$ of biogas. The additional water in this case is just to flush the manure into the tank, but in the case of the pigs or chickens the manure needs dilution to spread the bacteria around or to reduce the concentration of ammonia and other harmful material produced in the digestion. The hydrogen in the methane gas is from the water, so some water is essential.

First, we will calculate the size of the tank. The amount of manure from 10 cows is 15 ft$^3$. We need 25% more water, so the total amount of manure and water is

$$15 \times 1.25 = 18.75 \text{ ft}^3 \qquad [8.1]$$

It was previously noted that if the tank is not heated, it can take 50 days for manure to digest. We add 18.75 ft$^3$ of manure and water each day. The size of the tank then is

$$18.75 \times 50 = 937.5 \text{ ft}^3 \qquad [8.2]$$

We should add 10% for gas collection — a space above the liquid where the methane accumulates:

$$937.5 \times 1.1 = 1031.25 \text{ ft}^3 \text{ (Total tank volume)} \qquad [8.3]$$

Calculating the amount of methane gas from the 10 cows is straightforward, using the last column of Table 8.5:

$$10 \times 50 = 500 \text{ ft}^3 \text{of biogas/day} \qquad [8.4]$$

**Table 8.5. Manure and Biogas Produced by One Animal**

| Animal | Manure produced by one animal (ft$^3$/day) | Additional water % of manure | Biogas produced ft$^3$/day |
|---|---|---|---|
| Cow | 1.5 | 25 | 50 |
| Pig | 0.2 | 200 | 7.8 |
| Chicken | 0.004 | 800 | 0.4 |

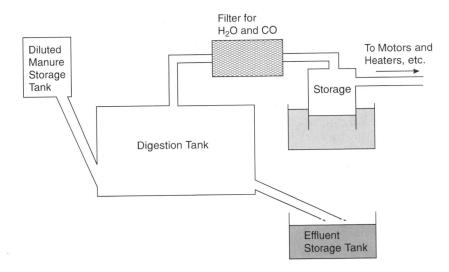

**Fig. 8.5. Methane digester — Overall system.**

A block diagram of the overall system is shown in Fig. 8.5. Holding tanks for both the manure and the effluent are provided in case one does not want to load the tank or fertilize the field each day. The filter removes hydrogen sulfide, generated from sulfur compounds in the manure. Hydrogen sulfide is corrosive and can damage engines. The filter consists of iron filings mixed with wood shavings. Limestone is also used in the filter to remove carbon dioxide produced in the gas.

If a faster reaction, which means a smaller tank (less retention time), is desired then the tank should be heated. One way to do this is to use the sun's energy, as in passive solar heating. Another way is to burn some of the biogas to heat water. The heated water is pumped through pipes inside the digester. At most, 20% of the biogas might be required for heating the mixture. The loss of gas in heating needs to be balanced with the savings in building a smaller tank.

## WHAT TO DO WITH THE BIOGAS

Table 8.6 shows estimates of the amount of biogas required to run certain appliances [5]. A mantle, as in a "Coleman" lantern, is used when gas is burned for illumination. The amount of biogas required by a family is estimated in Table 8.7. We assume two burners are on for 2 hr each and four mantles are on for 4 hr each. The refrigerator and the engine are both on for 4 hr. We see that the 10 cows will just suffice (see Eq. [8.4]). The

**Table 8.6. Quantities of Biogas Used by Various Appliances**

| Appliance | $ft^3/hr$ |
|---|---|
| Cooking (one large burner) | 20 |
| Cooking (one small burner) | 10 |
| Lighting (one mantle) | 3 |
| Refrigerator | 20 |
| Engine (5 hp) | 80 |

*Note.* 1 $ft^3 = 28.3$ liters.

**Table 8.7. Total Biogas Use by a Family**

| Activity/appliance | $ft^3/day$ |
|---|---|
| Cooking | 8 |
| Lighting | 16 |
| Refrigeration | 80 |
| Engine | 320 |
| Total | 496 |

amount of biogas generated by each cow can vary, so the numbers shown are not precise but are reasonable estimates for the United States.

In principle, any gasoline or diesel engine can be easily adopted to run on biogas. Actually, biogas is only 60% methane, so one needs to use more biogas than natural gas, which is nearly pure methane. People have driven tractors, even automobiles, using biogas, but a large storage tank is needed — 1 gallon of gasoline is replaced by 213 $ft^3$ of biogas. Biogas, of course, can be compressed into a smaller volume, but the compressor takes energy. Stationary engines are the most practical. Electric generators could be driven by these engines and a large farm could sell electricity back to the utility.

# APPROPRIATENESS OF METHANE DIGESTERS

Methane digesters have been built and used successfully both in the United States and in the Third World. Like many energy-producing projects, their value depends on the cost of fuel. If oil prices increase, then the biogas is more valuable. Also if a farmer has a great deal of manure to dispose of, then digesters look more attractive — to reduce odor if nothing else. In fact, though, methane digesters are not in common use, either in the United States or in the Third World. It is not clear why. Perhaps people do not know about them. Perhaps the collection of manure from fields is too difficult.

An issue not addressed here is the alternative use of the manure if a digester is not built. In the United States it would probably be put on a compost heap and then spread on the fields. The digester makes the handling easier and kills most of the germs, so it is beneficial. In parts of the Third World, the manure may be collected by the very poor and used as fuel, so use of a digester may hurt the very poor and help the more wealthy farmers. On the other hand, dried manure is not a good fuel: Its smoke irritates the eyes and can cause blindness. The issues of who benefits when a new technology such as this is introduced, are complex but cannot be ignored. We consider such questions in the second part of this book.

# REFERENCES

1. Chawla, O. P. (1986, January). *Advances in Biogas Technolgy*. Indian Council of Agricultural Research, New Delhi.

   *Much scientific data but also pictures and plans of working systems; gives data applicable to Third World animals.*

2. Kristoferson, L. A., and Bokalders, V. (1986). *Renewable Energy Technologies.* Pergamon, Oxford, UK.

3. Mother Earth News (1983). *Mother's Energy Efficiency Book,* pp. 148–167. Mother Earth News, Henderson, NC.

   *It includes plans for a model digester, a reasonable project, and a backyard digester — made from 55-gallon tanks. Chinese digesters are also described; readable and doable.*

4. National Academy of Science (1977). *Methane generation from human, animal, and agricultural wastes* (Report of ad hoc Advisory Committee on Technology Innovation). National Academy of Sciences, Washington, DC.

   *Thorough, readable, long bibliography; the examples are from India.*

5. Palmer, D. G. (1981). *Biogas, Energy from Animal Waste.* U.S. Government Printing Office, Washington, DC 20402.

   *Well-written report, focuses on cattle farmers in the United States; provides costs and an economic analysis.*

# Problems and Projects

1. How many pigs would be needed to get the same amount of biogas that 10 cows give? How large should the tank be? Would the C/N mixture be suitable? If not, what do you recommend?

2. Design an energy self-sufficient chicken farm for four people. Estimate energy needs. How many chickens would be needed to produce the required biogas?

3. Estimate building costs for a methane digester for 10 cows. Then estimate the cost per cubic foot of gas generated. Compare this with the gas rate charged by the local utility ($0.62/ft$^3$ per dollar in Rhode Island).

4. One problem with methane digesters is collecting the manure. Do feasible alternatives to keeping the animals in pens exist?

5. Design an integrated farm in which the effluent is used to fertilize food crops for animals and so forth. Does the idea seem practical? What questions need to be answered before making a final design?

6. (This is a construction project.) Build a methane digester. Chicken manure is a good input. A 1-gallon glass or plastic bottle can be used as the tank. Be careful that the pressure does not cause an explosion, which would be very messy!

7. Plan and build, if possible, a methane digester like Javed's in the United States. Where would you get components? How would you do the construction?

# 9

·········

# AGRICULTURE

··················

## WHAT CONSTITUTES TECHNOLOGY IN AGRICULTURE

It is striking how unfamiliar agriculture is to people in the United States. In many parts of the world, most casual conversations sooner or later get around to how well crops are doing, how much rain has fallen, and related topics. Such is probably not true for most readers of this book, so it seems worthwhile to describe some of the technological choices facing farmers. The important choices are (i) crop selection, (ii) tool selection (including fertilizer and pesticides), (iii) planting methods, and (iv) reaping methods.

## Crop Selection

Two questions are important in selecting the crops to grow: What do we want from the farm, and what is available at the site? Separate varieties of the same plant differ in terms of nutritional value, flavor, ease of growth in particular situations, disease resistance, difficulty of shipping and storage, and other considerations. In fact, a criticism of modern agriculture is that varieties are chosen on the basis of harvesting and processing factors rather than on nutrition and flavor. We need to decide whether we are farming primarily for food or cash. Do we wish to specialize? We must also study the character of our property. Different crops grow better on different sites. Rice grows better than wheat in marshy ground. Potatoes will give good yields in a sandy soil, whereas celery needs a rich soil. A match must be made between what we want and what we have to work with.

## Tool Selection

When one thinks of agricultural tools, one probably thinks first of hoes, shovels, or, at the other extreme, big tractor-driven cultivators or harvesters. Tools that are just as important are fertilizers that enrich the soil or herbicides and insecticides that kill harmful plants and insects. Tools include simple hand-held diggers and scrapers, small machines propelled by a person, and complicated diesel-powered machinery. In the same way, one can use natural fertilizers and pesticides, inorganic chemicals, or exotic products of recombinant DNA processes. What considerations bear on the choice?

A primary consideration is productivity — crop yield per acre. At least in the United States, another consideration is yield per worker. How much more rice can be harvested if a new kind of seed is planted which uses fertilizer and irrigation more effectively? How many more cows can a single farmer milk and care for with automated equipment? Certain tools have negative effects also. Chemical pesticides can run off a field and into a pond, harming fish or contaminating people's water supply. Some chemicals can build up in soil, causing long-term damage. Tractors make it possible to plant and harvest large fields but they may damage the soil by compacting it.

## Planting Methods

In order to germinate, seeds need moisture. In order to grow, their roots must be able to reach into the ground. Some seeds germinate best when exposed to sunlight, but they risk drying out. Seeds can be planted by merely sprinkling them on the surface, which is called the "broadcast" method. Birds will eat many of the seeds sown this way, and some of the rest will not grow because they land on unsuitable surfaces. If the seeds are big and the farm is small, the farmer can poke a hole for each seed with a stick and plant individually. A third alternative, using machinery, is tilling.

Tilling is usually done by plowing, which involves cutting and turning the soil to create a soft dirt—mulch. Tilling is sometimes said (but experts are not sure) to destroy the roots and seeds of weeds lying in a field. Modern tilling involves hooking up large, heavy plows to tractors and pulling them across a field, leaving behind cut up soil. Seeds are then deposited in shallow rows in the tilled soil. Deep plowing has disadvantages. First, it turns the topsoil into the subsoil, thereby "hiding" this most productive layer. Chemical fertilizer is then necessary to restore the soil fertility. Tilling also loosens the soil so that it may be eroded by the wind or rain. A human powered tilling machine developed for use in the Third World is shown in Fig. 9.1 [6].

Around 1970, people began using "low" or "minimum" tillage methods. These methods work only an inch or so of the top soil. Some studies show that the implementation of minimum tillage decreased the incidence of soil erosion to one-third the amount resulting from deep tilling. Minimum tillage systems still cause much soil erosion and are often chemical intensive (because herbicides are often used to kill weeds), so they are still not completely satisfactory.

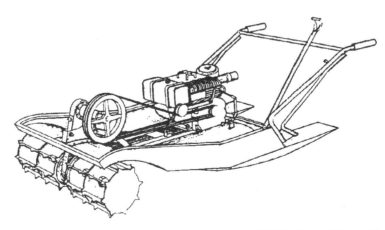

**Fig. 9.1. Small-scale rice tilling machine in the Philippines. (Reproduced with permission from IRRI [6].)**

## *Reaping*

Reaping is harvesting. Fruits and vegetables in the United States used to be picked by hand, often by migrant workers. Now some crops (e.g., tomatoes and grapes) are picked by machines which gently vibrate the branches so that ripe fruit falls into the machine. Machines can work 24 hours a day during the harvesting season and can collect more fruit than most workers because they can reach the less accessible parts of plants or trees.

Commercial grain farms harvest grain using combines. A combine is a large agricultural machine that rolls along a grain field and cuts, heads (takes off the top of the grain), threshes (beats the grain head in order to knock the seeds out), and cleans the grain. Combines are often huge (as large as a small house), driven from a cockpit-like cubicle. Combines range both in price and in size and can cost as much as a house. In their control cabs, some farmers install air-conditioning, stereos, and even televisions, telephones, and hot pots. The amount of time saved during harvesting by using a combine is tremendous, as is the amount of fuel used by some of these machines. The speed of harvest allows its completion during favorable weather conditions. However, these machines also compact the soil, so tillage is essential the following year.

Harvesting grain with a sickle or scythe is not fun but it was the only method known before the development of harvesting machinery. Hand harvesting involves cutting the grain with a large, curved knife, bundling it, heading it, threshing by hand, and then cleaning it by hand. It may take many hours to harvest an acre using only hand tools.

Intermediate methods of harvesting grain involve ox, horse, or hand-driven machines. These machines reap the grain and transport it to other machines; which are also driven by animal or human power, that thresh and clean it. Some of these small-scale machines are described by the International Rice Research Institute [6] (Fig. 9.1).

Now that we have an overview of how technology fits into farming, we examine an approach to Appropriate Technology farming.

# SMALL-SCALE FARMING

Small-scale farming can be carried out in many places and in many ways; the focus in this chapter is on possibilities in the United States. Yields can be high and the costs low. In essence, efficient small-scale farming is the substitution of knowledge and intelligence for the machinery and chemicals used in larger commercial farms. Small-scale farmers must know much about soil and plants and they must be able to solve unfamiliar problems, which are sure to arise. Stadfield [9] describes both the joys and the troubles of a small-scale farmer 50 years ago.

We now estimate the cost, in dollars, time, and space, and the yield from a small-scale farm in the United States. The objective of the farm we will consider is to raise sufficient vegetables for a family of four. The farm will use only natural fertilizer, no chemical herbicides and a minimum of pesticides — only those that leave no residue. In this case, the cost will basically be only that of the seeds and plants — perhaps $100 — and the land or enclosure, plus pesticides, if needed — perhaps $20. We will assume that the land is already available, as are tools. Since the cash requirement is minimal, we are concened with how much space is required and how much time is needed for care.

Small-scale farming can be done basically anywhere land is available. A UNICEF handbook [11] describes what they call "mixed gardens," fruit trees and vegetables around homes in the Philippines, and how these can provide enough food for a family. Vegetables can also be grown in greenhouses. They can grow on rooftops. Hydroponics (growing plants in an inert medium such as sand with a nutrient solution) can be implemented either inside or outside a building. Our main example will be intensive gardening as might be done on a corner of a field or in a backyard (Fig. 9.2).

Organic gardening uses many techniques of small-scale farming but on a commercial scale. Vegetables are grown without chemical fertilizers, herbicides, or pesticides. Some consumers will pay a premium price for organically grown vegetables. The value of organically grown vegetables is being increasingly recognized [4].

**Fig. 9.2. Small-scale farm.**

## INTENSIVE GARDENING

The system to be described is based on "French intensive" gardening, used in Paris suburbs toward the end of the nineteenth century, and "biodynamic" gardening, used later throughout Europe. The central idea is to use raised beds 4 or 5 ft wide and plant intensively so all the ground is covered. The width chosen makes tending the plot easier. The raised bed, only 6 in. or so above the ground, catches the sunlight effectively and the height clearly demarcates the boundary. If all the ground is covered, the sunlight is captured effectively, and weeds have difficulty growing.

Beds are prepared by digging deeply, approximately 12 in., or so, but not below the topsoil. Compost (described later), is added to the soil and mixed in. A thin layer of compost can also be placed around the sides of the beds to keep the soil in place.

The best seeds can be found by experimenting, asking people with similar soil and farming objectives, or by contacting (in the United States) the local Cooperative Extension Service. Some varieties are more pest and disease resistant than others. Planting several varieties of the same vegetable also lowers the risk of disease spread. One can save seeds from successful plants from the previous year but offspring of hybrid seeds may not grow back to be the same as the parents.

"Companion planting" is a sensible option. Several different vegetables, which will not compete, are planted in the same plot. A standard combination is corn, beans, and squash. The squash hugs the ground, the corn grows vertically, and the beans fill up the intermediate space. Some combinations reduce pests. Marigolds, in particular, are protectors of many plants. Garden books [7] give advice on which plant combinations are optimum. In Africa, where too much sun can be a problem, corn is planted among trees. The UNICEF handbook [11] also suggests planting trees with vegetables. Legumes, such as beans and peas, add nitrogen compounds to the soil and these compounds help nourish other plants.

Fertilizer does produce results. One can buy it or use compost or manure. Compost is decayed vegetable matter. It can be made in many ways. A pile of leaves will turn into compost within a year. The process is hastened by turning the pile over every month or so. Table scraps from the kitchen and other vegetable wastes can be added. Manure is a good source of nutrients and speeds the decay of the other material. Manure can be spread directly on the garden but raw manure is not good for a garden since it takes nutrients out as it decays. One should let the manure rot, either in the compost pile or by itself, before putting it on the garden. Of course, the effluent from a methane digester is another good fertilizer.

Pest control can be a problem, especially if one does not want to use chemical treatments. Pyrethrum and rotenone are two insecticides made from plants (pyrethrum is from a kind of chrysanthemum) which are not harmful to people, animals, or birds. Both decompose in the soil. If one prefers not to use these pesticides, one can try certain biological techniques. Birds eat insects, so by attracting birds one will reduce the insect population. Ladybugs and praying mantises also eat insect pests. If neither ladybugs nor praying mantises are attracted naturally to a garden, both may be purchased. Large insects can, of course, be picked off by hand, but the process often seems never-ending. "Traps," which are plants favored by pests, such as nasturtium, can also help by distracting insects from the actual crops. Another kind of trap is a small container of beer which attracts caterpillars away from squash plants. Covering the plants with a light cloth is another way of keeping insects away, but the cloth must be removed when the plants are ready for pollination.

Overall, perhaps as good a pest control "system" as any is to be willing to accept some blemishes in one's fruits and vegetables — being willing to just cut out the insect holes or damaged parts before eating. Another system is to plant thoughtfully. If several different varieties of a crop are planted, then it is likely that not all will be lost to disease or insects. Also, if plants of the same variety are kept apart, it will be harder for problems to spread. For the same reason, it is probably best to "rotate" crops, that is, to avoid planting the same crop in the same spot in consecutive years.

Care consists mostly of irrigating and fighting insects and weeds. Dense plantings crowd out most of the weeds. Drip irrigation systems, in which the water drips out of hoses along the surface of the ground, waste less water than sprinklers or massive application (dousing from a bucket) but require more precise operation.

Intensive gardens are harvested by hand. Actually, the biggest problem (as anyone who has had such a garden knows) is making use of the produce when it ripens. Some crops, such as beans, can be dried. Others, such as tomatoes, can be canned. Most crops can be frozen, but the freezing and storage uses much energy. Cabbage, potatoes, and turnips will keep in a cool place. In principle, one could barter fresh vegetables for other necessities, but the "barter value" of vegetables at harvest season is limited.

## Inputs to and Yields from Intensive Gardens

Now we consider how much food we can get from a small garden. The yields in Table 9.1 are based on the New Alchemy work [1, 7], the UNICEF handbook [11], and personal experience. Table 9.1 shows, for example, that 0.22 lbs of beans can be harvested from one square foot of garden, and that amount of beans corresponds to about 1.6 portions for an adult. To estimate the amount of space needed to feed one person for 1 week we need to hypothesize a menu, which is provided in Table 9.2. (This person eats potatoes twice a week for breakfast, but otherwise does not eat vegetables at breakfast.) Knowing the number of portions eaten per week and the space required per portion (obtainable from the last column of Table 9.1) we can find the space needed to raise a week's worth of vegetables. The space per person for a week's vegetables is 12.44 ft$^2$.

Thus, an estimate of the total space needed in a garden for four people for a year is

$$12.44 \times 4 \times 52 = 2587 \text{ ft}^2$$

This corresponds to a growing space 26 × 100 ft (approximately one-sixth the size of a football field, or 1/16 of an acre). If one used intensive gardens 5 ft wide and 20 ft long, 26 gardens would be required. The total space requirement would have to be larger, perhaps by 50%, to accommodate the aisles between the growing areas.

Now we will calculate how much time it would take to raise these vegetables. The time input consists of bed preparation, planting, irrigation, insect removal, harvesting, and a few extra tasks. Most of these times can be estimated without difficulty. The irrigation time depends on the method used and how much of the time waiting for the water to soak in can be used

### Table 9.1. Yields of Vegetables in Intensive Gardening

| Crop | Yield per square foot (lbs) | Portions per square foot (individual) |
|------|-----------------------------|----------------------------------------|
| Bean | 0.22 | 1.6 |
| Broccoli | 0.38 | 2.1 |
| Cabbage | 1.70 | 21.0 |
| Celery | 0.19 | 1.5 |
| Eggplant | 0.37 | 1.6 |
| Lettuce | 0.41 | 2.4 |
| Onions | 0.57 | 2.6 |
| Potatoes | 2.30 | 12.8 |
| Spinach | 0.19 | 0.9 |
| Tomatoes | 1.00 | 3.0 |
| Turnips | 0.19 | 0.9 |

### Table 9.2. Space Required to Feed One Person for 1 Week

| Crop | Portions per week | Space needed (ft$^2$) |
|---|---|---|
| Beans | 4 | 2.50 |
| Broccoli | 3 | 1.40 |
| Cabbage | 2 | 0.09 |
| Celery | 2 | 1.30 |
| Eggplant | 2 | 1.30 |
| Lettuce | 4 | 1.70 |
| Potatoes | 7 | 0.55 |
| Spinach | 2 | 2.30 |
| Tomatoes | 4 | 1.30 |
| Total space | | 12.44 |

productively elsewhere. A reasonable but high estimate for irrigation might be 45 min a week during the growing season. Insect fighting would probably average 1 hour a week. Table 9.3 provides the time estimates. The 290 hours shown in Table 9.3 are done during May–September for about 60 hours a month. An experienced worker could probably reduce this time by one-third, to about 40 hours a month.

As described earlier, the annual cash inputs are a minimal — $50–90 a year. Small tools (hoe, rake, hose, shovel, wheelbarrow, buckets, and baskets) are needed, but these should last 10 years; the annual cost is perhaps $25 a year. The land itself has not been included in these cost calculations, nor has the cost or availability of water for irrigation.

### Table 9.3. Time Estimates for an Intensive Garden

| Activity | hr |
|---|---|
| Bed preparation | 25.0 |
| Planting | 12.5 |
| Irrigation | 90.0 |
| Insect removal | 125.0 |
| Harvesting | 12.5 |
| Odds and ends | 25.0 |
| Total hours | 290.0 |

## Comparison with Commercial Farms

We should compare the costs and yields of our intensive garden with those of a commercial farm. A commercial farm does not have a much greater yield (at most twice as much) but requires significantly less labor—about 5% of the labor if automated harvesting is used. The trade-off is between labor costs, on the one hand, and machinery and chemical costs, on the other hand. Organic farmers and intensive gardeners point out, however, that their methods preserve, and even improve, the soil, whereas commercial farming causes erosion, chemical buildup in the soil, and compaction of the soil.

# GREENHOUSES

A small solar-heated greenhouse (Fig. 9.3) can produce vegetables year-round in Northern climates, although winter crops are probably restricted to leafy vegetables such as lettuce, spinach, and (for the experienced) celery. Tomatoes will grow in fall and spring—the limiting factor seems to be the amount of sunlight available. Reports from the New Alchemy Institute [1, 7] are good sources of advice on greenhouse vegetable gardening. A greenhouse does not have to shut down in the summer, but one has to be careful about overheating. Soils need more attention in a greenhouse than outside because propagation is more intensive and the ecosystem is probably incomplete. At best, yields per square foot in a greenhouse can be

**Fig. 9.3. Owner-built vegetable greenhouse.**

the same as those in an intensive garden. Insect pests can be more of a problem because natural controls are absent and the compactness facilitates movement from plant to plant.

## ROOFTOP GARDENS AND HYDROPONICS

People have succeeded in growing vegetables on roofs in cities. Major problems are drying winds and the baking sun. Windscreens and frequent irrigation are practical solutions to these problems. Of course, one must be careful to verify that the building itself can withstand the weight of the soil and water. A compost pile may not be feasible, so other ways of conditioning the soil must be provided. Rooftop gardening is still in the experimental stage, but it could be fun to try.

Hydroponics is less of an experiment. Garden supply stores sell racks in which trays of plants can be held connected by tubes carrying the nutrient solution. On the other hand, one could build one's own system. A hydroponic system is shown in Fig. 9.4.

The container containing the nutrient solution is raised several times a day, as shown in Fig. 9.4, to flood the growing trays. When the container is lowered, the solution drains back. The growing medium can be sand, gravel, or plastic particles. The nutrient solution can be chemicals or natural fertilizers, such as water from a fish tank.

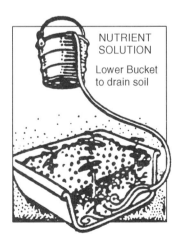

NUTRIENT SOLUTION

Lower Bucket to drain soil

**Fig. 9.4. Small hydroponic system. (Reproduced with permission from *The Bountiful Greenhouse* [8].)**

More involved hydroponic systems use rafts of plants floating in fertile water, such as in a fish pool. An advantage of hydroponics is that one has control over the growing medium and the nutrient solution, although use of fish pond water reduces that control. Shane Smith [8] describes hydroponics more fully.

# ANIMALS

Nutritionists disagree about whether animal protein is essential in a person's diet, but for most people life would be significantly different without it. The next chapter considers fish farming in detail. Here, some animal production possibilities for a small-scale farmer are mentioned.

Chickens are easy to raise and are probably the most productive animals per unit of area. With little care from a farmer, each hen will lay an egg approximately every other day. A dozen hens could keep a family of four well supplied for most of the year. Chickens will eat tomato bugs and other insects if allowed to run free, but they should also be fed grain and commercial supplements. An acre of corn will feed 25 chickens for a year. New chicks can be purchased or raised from fertilized eggs. In the latter case, both a rooster and an incubator are needed. It is not hard to build an incubator: A 100-W lightbulb is an adequate heat source. If hens are kept in pens, which makes finding the eggs much easier, about 4 ft$^2$ per hen is necessary. A 7 × 7 ft area would take care of 12 hens. Stadfield [9] in his book about the family farm, points out that chickens are productive but "nobody loved them very much."

Rabbits are much more lovable — so much so that some people refuse to slaughter them. Rabbits, as folklore states, are prolific. One pair of rabbits will produce four litters a year of approximately six rabbits each. At only 6 month of age, a rabbit is ready to mate, so it is easy to have many rabbits quickly. They eat hay, soy beans, and vegetable scraps. Four rabbits need a space 3 × 4 ft.

Goats are recommended as "the domesticated animal of choice" for small spaces [10]. Goats give milk and eat almost anything organic, but they prefer leaves and grass. Their meat is also edible. In some parts of the world, when one orders mutton in a restaurant one gets goat meat. A goat will give three quarts of milk per day for 10 months if fed well and bred annually. Goat's cheese is as well regarded as is the milk. Goats are also attractive and friendly pets. They breed easily, so it is easy to generate a big flock, even if one's primary objective is to get milk (Fig. 9.5)

Ducks provide eggs and nutritious meat (not all at once, of course). They can be grown in coordination with fish ponds, forage on their own, or

**Fig. 9.5. Goats.**

can be fed commercial food. A pond of some kind is a necessity. Ducks and geese are useful for weed control but, because they eat desired vegetables as well, they should be used to clean a plot before planting or after harvesting. Ducks eat slugs and insects so they are useful for pest control, particularly in the fall when insects are preparing to dig into the soil for the winter [3].

Cattle supply both milk and meat. Dairy cows are labor-intensive because they must be milked each day. Cattle eat hay and grass, which is usually supplemented by grain and commercial food. The cost of milk from one's own cow is about 25¢ a quart. In comparison to chickens, cows are lovable but require a greater commitment.

Bees are also not difficult to raise, although one needs special equipment — both hives and protective gear for handling. One usually begins by buying a swarm of bees. Once the hive is set up, little care is needed. People even raise bees successfully in urban areas (Fig. 9.6).

## PRACTICALITY OF SMALL-SCALE FARMING

The previous calculations show that it is feasible to grow one's own vegetables on a small plot, and the economics are favorable, if one already owns the land. Corn seems to be the only grain that can be grown efficiently on a small scale (10 acres or less), at least in the northern parts of the United States. Growing one's own vegetables, though, involves heavy and tedious work. Of course, the food obtained is, fresh, and the grower knows

**Fig. 9.6. Beehives.**

what chemicals have been applied. Some people enjoy the gardening itself, which provides both the opportunity to work with plants and the challenge of producing good crops.

Community gardens in urban areas are an example of successful small-scale farming (Fig. 9.7). Vacant lots are often used. The soil is enriched with compost, perhaps from leaves donated by the parks department. The soil should be tested for the presence of heavy metals, particularly lead. City agencies or community organizations in many cities will give advice and even lend tools. Often the gardeners depend on the food grown for

**Fig. 9.7. Urban garden.**

their own meals, so productivity is high. In some U.S. cities, refugees from other parts of the world have been leaders in making these gardens successful, perhaps because they had agricultural backgrounds in their home  countries. The neighborhood benefits by having a vacant lot turned into a productive and attractive garden. Urban gardens are discussed in Chapter 14.

Raising money from small-scale farming seems to be difficult. As will be described in Chapter 14, a small producer does not fit well into the food distribution system in the United States. The relative cost of labor compared to machinery makes large-scale commercial farming profitable, with food prices even lower than the costs of small-scale farming. On the other hand, proponents of small-scale organic farms state that commercial farming has long-term deleterious effects and that the economics are only favorable now because that long-term cost is not being recognized.

## THE GREEN REVOLUTION

The "Green Revolution," which took place in India, Pakistan, and China during 1960–1966, is a combination of successful plant genetics and economic policies. Wheat and rice plants were bred to grow strong, short stocks with high yield and shorter maturation periods. Governments established national production campaigns providing seed, fertilizer, and helpful credit policies, and they guaranteed harvest prices equal to those of the international market. The result was that farms produced two or three times their previous yields through multiple cropping. This major break-through led these overpopulated and underproductive nations to a point of self-sufficiency in cereal production.

Much has been written about the Green Revolution [2, 3]. Use of special seeds and fertilizer creates a dependence on external suppliers and on generating cash. The social consequences of bringing farmers into the cash economy has been different for men than for women and has tended to make the situation of the "poorest of the poor" worse, as will be described in Chapter 13. Another observation about the Green Revolution is that its early successes have not been duplicated. For example, successful crops for arid lands, as in southern Africa, have not been found. Proponents of the Green Revolution say the revolution is not yet complete.

## IS HIGH TECHNOLOGY APPROPRIATE?

Farming is an activity in which it is possible to compare directly the high-technology approach with the intermediate-technology approach. High-technology farming uses machinery and chemicals extensively as well as specially bred animals and plants. An automated egg farm has individual cages for each hen about the same size as the hen. Conveyor belts provide feed in front of the hen and collect eggs behind the hen. Manure drops through the bottom of the cages. The feed is a specially prepared mixture, including hormones for high productivity. Another example of high-technology farming is a tomato plantation which uses machinery for planting and harvesting. The harvesting machine covers several rows at once, shaking the plants to remove the ripe tomatoes. It works 24 hours a day during the harvesting season. The variety of tomato is selected for resistance to bruising as well as yield (flavor may be sacrificed). Chemical fertilizer is used extensively, as are insecticides to kill insects and herbicides to kill weeds. Output per acre is higher than that in a traditional farm, and output per person is even higher. No unskilled laborers are needed and comparatively few other workers are needed. The high yield and the large capital costs mean that the farm must have guaranteed markets — either supermarkets or processors.

If the yield is high, why might one be concerned about high-technology agriculture? In the United States high-technology agriculture has reduced the need for farm workers, creating social problems, although neither the work itself nor the living conditions for migrant farm workers were desirable. Actually, pressure from farm workers' labor unions may have been as much an inducement for mechanization as the economics involved. It is easier to maintain and operate machinery than to manage large numbers of workers, and working conditions are less of a concern. In any case, mechanized farming has taken jobs away from many people.

Another problem is damage to the soil, as mentioned previously. Damage results partly from chemical residues and partly from the weight of the machinery moving down the rows. The open planting needed to give the machines space to move allows erosion. Some of the chemical residues, including hormones fed to animals, end up in the meat with unknown health effects. Yearly use of the same crop variety may be dangerous in the long run because diseases and pests matched to that variety may multiply.

The effectiveness of high-technology farming in the Third World is even more questionable. In one Third World country, a chicken farmer using advanced equipment and facilities became dependent on foreign sources for feed as well as for spare parts. When foreign exchange became tight, the farmer ran into trouble because he could not purchase feed nor maintain his equipment. Local markets in the Third World are another aspect of the problem because they usually are too small to consume all that a mechanized farm can produce, so part of the output must be exported. The result is that the mechanized farm is closely tied to foreign businesses for both supplies and markets. Unexpected changes in either costs or sales can be disastrous. Even mechanized poultry and egg farmers in the United

States have found themselves in serious trouble when fuel prices have increased.

High-technology agriculture can be appropriate where labor is expensive or difficult to manage and the farm is integrated into a system of suppliers (chiefly chemicals and machinery) and markets. The high-technology farm also needs provisions for maintaining the quality of the soil and market demand for the vegetable or animal variety. When all the conditions just stated are not met (critics would say the conditions cannot be met in the long run) then a less technologically intensive agriculture may make better sense. The long-term success of mechanized farming — whether it can be sustained — must be considered carefully, although one could argue that the size of the commitment that has already been made to high-technology agriculture, especially in the United States (which feeds much of the world) implies that a change away from high-technology agriculture is not feasible.

# REFERENCES

1. Armstrong, C. (1981). Indoor gardening. *J. New Alchemists*, **7**. The New Alchemy Institute, P.O. Box 432, Woods Hole, MA 02543.

   *Describes gardening in their bioshelter, basically a greenhouse with a much more complete ecosystem than most greenhouse; gives hints and compares varieties of seeds.*

2. Borlaugh, N. (1986, Summer). Accelerating agricultural research and production in the Third World. *Agric. Hum. Values* **3**(3).

   *An analytical approach to stimulating agricultural development in African countries. Borlaugh is a leader in the Green Revolution.*

3. Jesiolowski Cebenko, J. (1977, July/August). Weeds & pests? Get ducks and geese! *Organic Gardening* **44**(6), 26–33.

   *Describes varieties of ducks and geese, plus provides advice on care; gives sources for the birds.*

4. Greener greens; The truth about organic food (1988, January). *Consumer Rep.* **63**(1), 12–18.

5. Critchfield, R. (1983). *Villages.* Anchor, New York.

   *A personal-level look at specific villages, including how food is grown, in developing nations and the village situations; describes the results of the Green Revolution.*

6. Annual reports, The International Rice Research Institute, P.O. Box 933, Manila, Philippines.

   *IRRI is best known for its rice seedbank but has been active in small-scale agricultural tool development. The 1984 and 1985 Annual Reports are especially helpful.*

7. Maingay, H. (1977). Intensive vegetable production. *J. New Alchemists* **4**.

   *Very helpful discussion of intensive gardening, with a full bibliography. Start here for good ideas and encouragement. Further discussion, with an energy analysis, is given in the following volume.*

8. Smith, S. (1982). *The Bountiful Greenhouse*, John Muir Publications, Santa Fe, NM.

   *Hydroponics are described in many greenhouse guides. This one does it well. Other tips for inside vegetable growing included; oriented toward user, not the scientist.*

9. Stadfield, C. K. (1972). *From the Land and Back — What Life Was like on a Family Farm and How Technology Changed It*. Scribners, New York.

   *The subtitle says it all: Life was tough but satisfying and as technology made the work easier, the economics got worse.*

10. Thomson, C. (1997/1998, December/January). The small herd. *Rural N. Engl.* **86–87**.

    *This issue has several articles about raising goats; certainly enough information so the reader can decide whether to do it.*

11. UNICEF (1982). *The UNICEF Home Gardens Handbook*. United Nations, New York.

    *Focuses on the Philippines; meant "for people promoting mixed gardening in the humid tropics," so it is partly horticulture and partly advice to aid workers.*

12. U.S. Department of Agriculture (1978). *Living on a Few Acres*. In *The Yearbook of Agriculture 1978*. U.S. Department of Agriculture, Washington, DC.

    *I received mine free from my U.S. senator. I do not know if a recent edition is available. Straightforward, friendly articles — how to choose a site, what fruits and vegetables to grow, and what animals to raise. Inspirational!*

## PROBLEMS AND PROJECTS

1. Design a garden to raise sufficient vegetables to feed a family or group of four people. Consider the kinds of vegetables to be raised, alternative ways of having vegetables available in the winter, amount of land needed, amount and kind of fertilizer, method of insect and weed control, and tools for planting, cultivating, and harvesting.

2. Design a greenhouse garden to raise sufficient vegetables to feed a family or group of four people. Design the system, choosing types of vegetables, size of greenhouse, and so forth.

3. Design a hydroponics system to grown food in a particular building, for example, a home. How much food could be obtained? What are the costs? Does the idea seem worthwhile? Try to build a model to test your system.

4. Consider the qualities of a "perfect crop" (i) in the United States and (ii) in a less industrialized country. List the attributes you would seek. Actual perfection is difficult to attain, as we all know, so rank the attributes in order of importance.

5. Assume you are the minister of agriculture for a less industrialized country, and a person from an international aid organization comes to you with a new plant or variety of existing plant. The plant in question has yielded four times as much grain as the current traditional crop of your country and the visitor urges you to replace that traditional crop with the new one. What specific questions need to be answered in evaluating the visitor's proposal?

6. How could you make a small-scale farm financially successful in your locality? Do specific crops exist in high demand? Could you board horses? Could you take in vacationers? Can the produce be sold without tying up an essential person?

7. (This project and the next must actually be performed.) Grow a food crop, such as lettuce, in an urban, winter setting. Experiment with techniques, such as use of a window sill, use of artificial light using chemical fertilizer, and using natural fertilizer. Estimate the cost of the crop grown in dollars per serving. Discuss why you chose the crop you did.

8. Experiment with making compost. Some parameters to be considered are type of organic matter used, number of times the pile is turned and additives. Compare the time taken for decomposition for the various parameters tested. Measure the temperature rise (to kill harmful germs the temperature must reach higher than 130°F).

# *10*

· · · · · · · · · ·

# AQUACULTURE

· · · · · · · · · · · · · · · · · · · ·

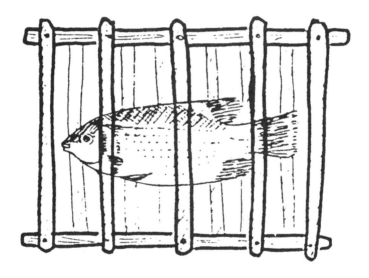

## DEFINITION

*Aquaculture* is fish farming. It can be a good use for marshy or infertile land. A farmer, either in the United States or in the Third World, can earn cash from aquaculture or can simply feed a family. This chapter is chiefly about aquaculture in ponds in the United States, although aquaculture can be practiced on ocean shores and bays [7]. Some people enjoy eating fish and some people, not necessarily the same ones, enjoy catching them. Fish grow rapidly even in a small pond. One can get more protein per acre growing fish than raising most animals and certainly much more than raising cattle. Depleted soil is made fertile while it is the bottom of a fish pond. Ponds can be attractive and fun.

Fish farming can be done in many ways. A stream can be dammed, an area can be dug out and dams built on four sides, or a concrete pool can be built. Local fish can be raised or selected breeds can be stocked. The fish can be allowed to forage for their own food, eating the vegetation and animals growing wild in the pond. On the other hand, yields can be increased by feeding the fish directly or by fertilizing the water so small plants and animals multiply. If the density of fish in the pond is kept small culture is easy, but if many fish are living in a small volume, loss of oxygen in the water, disease, and overcrowding must be considered.

The basic science of raising fish is not difficult; many people raise tropical fish successfully at home and fish farming is just an extension. Fish need water, food, and oxygen. Small fish, newly hatched, need protection from big ones. In fact, fish can grow and breed so successfully that they can overpopulate their pond or pool, resulting in many fish too small to be of use.

## KINDS OF FISH

Many different kinds of fish have been raised artificially. Perhaps the most common choices in the United States are catfish or a combination of bass and sunfish (Fig. 10.1). Catfish farming is a big industry in southern states, with most of the fish sold commercially. Catfish will survive even in fairly dirty water. Bass and sunfish coexist well in ponds. Both bass and sunfish are good for eating and desired for recreational fishing. One reason a bass–sunfish combination makes sense is that the bass eat small sunfish, which keeps that population under control — without the bass, there would be too many tiny sunfish. Other choices for fish are mentioned at the end of this chapter.

## FEEDING

There are at least three possible approaches to feeding: (i) do nothing, relying on the natural fertility of the pond; (ii) use fertilizer to grow plankton (small plants and animals), and (iii) put fish food into the pond.

A basic food for fish is plankton. The young of most fish, including catfish, bass, and sunfish, eat only plankton, and mature catfish and sunfish can live on plankton alone. Plant plankton, also called *phytoplankton*,

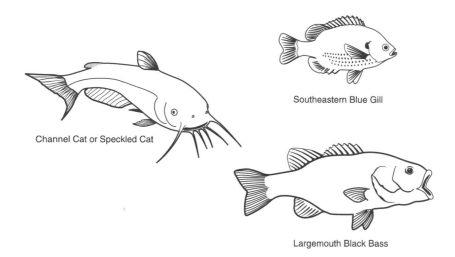

Channel Cat or Speckled Cat

Southeastern Blue Gill

Largemouth Black Bass

**Fig. 10.1. Types of fish.**

produces the algae blooms which turn pond water green. Animal plankton, *zooplankton*, consists of tiny shrimps, insects, worms, and so forth. Zooplankton eat phytoplankton. To grow larger numbers of fish, the grower can fertilize either the phytoplankton or the zooplankton in a typical pond. Inexpensive inorganic fertilizers, such as the ones used on backyard lawns, feed the phytoplankton, but zooplankton need a more complex fertilizer, such as animal manure, compost, or commercial plant protein meals. (Some people, however, object to putting manure into a pond.)

One feeding strategy for a new pond is to use both commercial backyard fertilizer and manure or compost for the first few weeks, so the zooplankton have something to eat until the phytoplankton is established. After the pond has been established for several weeks, plant fertilizer suffices. By this time the water will look murky because of the phytoplankton. Incidentally, in fish farming, in which one wants plankton in the water, murky water is desired, whereas tropical fish hobbyists pride themselves on the clarity of the tank water.

Supplemental feeding with commercial fish food (feeding the fish directly) can raise the productivity of a natural pond by a factor of approximately four. It is estimated that each pound of harvested catfish requires about 1.5 lbs of food. Of course, this amount of food is not given all at once — just like people, larger fish need more food. Different fish do best on different formulas. Besides commercial fish food, many common products are appropriate: wheat, rice, corn, insects and larvae, table scraps, and so forth. An efficient way to learn what fish will eat is to experiment.

## HOW MUCH FERTILIZER TO USE

The amount of fertilizer needed depends on the fertility of the soil at the bottom of the pond and the nutrients in the water. Each spring one might have to fertilize a typical pond five times, putting in 8 lbs of nitrogen and 10 lbs of phosphate per acre each time. This much nitrogen and phosphate corresponds to 50 lbs of 16-20-0 commercial fertilizer or 100 lbs of 8-10-0. One might instead use 180 lbs of chicken manure, 800 lbs of cow manure, or 500 lbs of grain (all per acre). The commercial fertilizer can be placed in fixed or floating containers in which it can seep away gradually. The manure is best placed in a porous bag so the water can flow through and dissolve the manure. The grain can simply be thrown into the pond.

A simple test indicates if the amount of fertilizer in the pond is proper; that is, if the amount of phytoplankton is proper. To test this, one sticks one's arm into the water. If the hand is invisible when the elbow is just at the surface, the water probably is sufficiently fertile. If the hand becomes invisible with the elbow below the surface, the water is not fertile enough. If the hand is invisible with the elbow above the surface, the water is too turbid, which may cause problems of oxygen depletion. The technical term describing water clarity is turbidity — turbid water is not clear.

A more sophisticated device for measuring water turbidity is the *Secchi* disk (Fig. 10.2). The disk is 8 in. in diameter. The disk is lowered into the water on a string until it disappears. One expert suggests the optimum Secchi distance, the distance before it disappears, is 43 cm or about 17 in. A good range is 12–24 in.

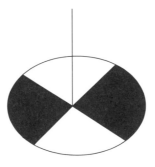

**Fig. 10.2. Secchi disk.**

# TAKING CARE OF A FISH POND

Probably the biggest danger to fish life is oxygen depletion in the water. Fish are like mammals: They breathe oxygen and emit carbon dioxide. Fish get their oxygen from that dissolved in the water and if the amount dissolved drops too low, fish die — people who have put too many tropical fish in a tank without an aerator know this. During daylight, plants take in carbon dioxide and emit oxygen, which is why the world has not run out of oxygen. At night, however, the plants do not produce oxygen and, in fact, use it. The most dangerous time for the fish is dawn, before the plants, chiefly the phytoplankton, have begun producing oxygen. Hot weather is more dangerous than cold because less oxygen can be dissolved in hot water. Oxygen depletion is a major pond risk; all the fish in a pond can die overnight if it occurs.

Phytoplankton affects the amount of dissolved oxygen in another way. More phytoplankton makes the water less transparent, so the sun's energy penetrates to a lesser depth. This lesser penetration is a problem because oxygen is not produced at depths where the sun's energy does not reach. In fact, one way to destroy bottom-growing weeds, which can make fish harvesting difficult, is to encourage dense plankton growth (plankton absorbs sunlight). Maintaining the proper amount of plankton in a pond is probably the major pond management problem.

The oxygen content of the water can be measured using commercial kits, but these are moderately expensive (approximately $50). The Secchi disk is a simple tool. If the disk disappears at less than an 8-in. depth, the grower should quickly increase the oxygen content. Another indication of oxygen deficiency is the fish themselves. If at dawn they are near the surface, gulping for air, the pond water is low in oxygen. One expert suggests watching for the presence of flocks of birds, which are attracted by fish at the surface.

What can be done if the dissolved oxygen level is low? Splashing the water, using either paddles or a propeller driven by a motor, will expose more water to air and allow more oxygen to be dissolved. In China, people stir and beat the water with poles. Releasing bottom water from a pool removes decaying vegetable matter as well as water with low oxygen content. If the replacement water is splashed in, it will pick up enough oxygen to increase the dissolved oxygen content significantly. It is best to design the pond so the inlet water splashes into the pond (Fig. 10.3). Of course, such design advice will not be helpful if the pond is already built.

In addition to oxygen depletion, other negative things can happen to a fish pond. Predators, such as snapping turtles, snakes, cormorants, and even alligators, can steal fish. Pesticides can get into the pond through the

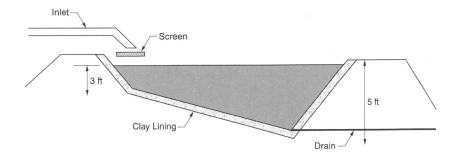

**Fig. 10.3. Side view of fish pond.**

incoming water or even through the air, if a neighboring farmer used crop dusting. Disease can strike. One recognizes disease by unhealthy-looking, slow-eating fish. The best advice when disease is suspected, at least in the United States, is to call the local fishery extension service or the state university.

## POND DESIGN AND CONSTRUCTION

Like other aspects of aquaculture, pond construction can be simple or complex. One can use an existing pond, build a concrete pool, or design something in between. Ponds can be small or large, and water can be obtained from streams, springs, or wells. Relying on rain is risky, as is relying on runoff from nearby raised ground—the water coming in may contain herbicides or pesticides.

The first step in pond design is probably to choose a location. The choice depends on many factors: the water source, how well the soil holds water, ease of bringing construction equipment in, other possible uses for the area, and so forth. In general, a good place to consider first is a low part of an unused field.

An essential early step is to check the required permits, especially if one is modifying "wetlands." The permit process can be long and tedious but if omitted can result in forced removal of the pond. Someone in the local town hall, county office, or equivalent should be called for advice.

For a fish pond in a small farm, the smallest practical area is about 1 acre. The fish population in a smaller pond will probably be too small to be stable. The Fish and Wildlife Service recommends 20-acre ponds for catfish. If the water will not freeze, 3 or 4 ft is a suitable depth; not much of the sun's energy will penetrate lower. In cold climates, 6 ft is probably needed so sufficient water will be available below the ice in winter.

Pond construction was discussed in Chapter 4. A pond can be made by digging out the earth or building walls, called levees, around the pond site. The soil in the pond bottom should be tested to make sure it will hold water; clay soils will. Otherwise the pond must be sealed. The levees can be made of earth but must contain sufficient clay to resist leakage. The walls should be built with a 2 : 1 slope, as shown in Fig. 10.3. A steeper slope may cave in; a more gradual one causes shallow areas where weeds can grow. Construction of a pond will be easiest with a bulldozer. Walls higher than 6 ft should be checked by a civil engineer. If the soil at the site leaks, a clay layer can be installed. Another alternative is polyvinyl sheets, which then are covered with earth. The New Alchemy Institute [6], recommends pig manure as a sealant.

The pond should be built with a drain so repairs can be made to the dam or so water contaminated by disease or pesticides can be removed. If the pond is made by excavation and is lower than the surrounding area, the drain needs a pump. The inlet water, as described previously should drop into the pond for aeration. An inlet screen may be necessary to keep out wild fish which may carry disease or may crowd out or eat the desired ones.

The bottom of the pond should be cleared of weeds before filling. Sunfish can hide in weeds, thus reducing the population-control effectiveness of the bass. Food pellets can be lost in weeds and be useless to the fish. When the weeds die and decay, they take oxygen from the water. Weeds on the bottom, alive or dead, make it difficult to use nets to catch fish, and weeds that grow through the water, above the surface, are a home for mosquito larvae. Even if the pond is cleared, some weeds will grow after it is full and these need to be removed. To prevent such aquatic weeds, there should be no shallow areas in the pond.

The water can be obtained from springs, wells, streams, or runoff. Springs, if available, are the best choice. Well water tends to be clean but usually has little dissolved oxygen and probably needs a pump. Part of a stream can be diverted to feed a pond, but it must be filtered to remove wild fish. Stream water, however, may bring in suspended dirt, thus increasing turbidity. At the same time, the water flowing back into the stream removes nutrients from the pond. Runoff water can be used and may bring nutrients into the pond, such as when the runoff comes from an animal yard. An adequate water supply must be ensured, however, and the water must not carry in poisons from the runoff surface.

An appealing idea is to integrate a fish pond into a chicken or pig farm by using the manure from the animals to fertilize the pond, perhaps by directing runoff from the animal areas to the pond. Some of the rich pond water could, in turn, be used to irrigate gardens.

A potential fish farmer can receive help from several places, including the books listed in the References. Information is available from the fishery extension service in most states or from local state universities. The U.S. Department of Agriculture program to encourage fish ponds has been curtailed, although advice is still available.

## STOCKING AND CATCHING THE FISH

It is possible to breed (*spawn*) fish oneself, but the process is somewhat difficult. Specialists recommend that the farmer obtain, *fry*, newly hatched fish, or *fingerlings*, finger-long young, from hatcheries. Representative prices are 20¢ each for catfish and bass and 05¢ for sunfish, although some federal hatcheries will give bass and sunfish away for free. Appropriate annual stocking rates per acre for fingerlings are shown in Table 10.1.

If one wants to breed one's own fish, one needs to set up places to do so. Catfish need spawning containers, such as 10-gallon milk containers. Small heaps of gravel will work for bass. The yield of adult fish is increased if the fry are moved to separate small ponds so they are not eaten by the big fish. Breeding is not impossible but one should learn about it from books or successful breeders before attempting to breed one's one fish.

Harvesting can be done by hook and line, by using nets, or by draining the pond. Of course, if one drains the pond one will have to dispose of many fish in a short time. In fishing a bass/sunfish pond one must maintain a 1 to 4 ratio of pounds of bass to pounds of sunfish or about 1 bass for 20 sunfish. As long as one maintains the correct proportion of fish one cannot overfish a small pond with a hook and line — the more fish caught the more food left for the others.

The yield from a fish pond is shown in Table 10.2 in pounds of fish per acre per month. In warm climates ponds can probably produce 12 months a year. In colder climates production is less than the numbers shown in Table 10.2 as the fish stop eating and breeding in the winter. Table 10.3 gives approximate prices of fertilizer and fish foods.

**Table 10.1. Number of Fish to be Stocked Annually**

| Type of fish | No care | Fertilizer | Feeding |
|---|---|---|---|
| Channel catfish | 75 | 400 | 2000 |
| Bass/sunfish | 50 bass/500 sunfish | 120 bass/1200 sunfish | 200 bass/2000 sunfish |

### Table 10.2. Estimated Pounds of Fish per Month per Acre of Pond

| Type of Fish | Yield in lbs/acre/month | | |
| --- | --- | --- | --- |
| | No care | Fertilizing | Feeding |
| Channel catfish | 10 | 37.5 | 200 |
| Bass and sunfish | 1.5 | 15 | 25 |

### Table 10.3. Estimates of Prices of Fertilizer and Fish Food

Inorganic fertilizer $15/10 lbs nitrogen

Catfish food $400/ton

Bass and sunfish food $475/ton

# YIELDS AND COSTS OF A SMALL FISH POND

Does fish farming really make sense for a small farmer? The following example considers the economics of a bass/sunfish pond in the northeast part of the United States. Catfish in the southern United States or carp in the Far East would produce higher yields.

In a 1-acre pond, when only fertilizer is added, about 15 lbs of fish could be harvested each month from March to November (see Table 10.2) and perhaps half that in the other months, about 150 lbs a year. If each person in a four-person household ate one-half a pound of fish a meal, the pond would supply fish for nearly two meals a week in the 8-month period, which would be a significant part of a family's diet.

How much do these fish dinners cost? The annual costs are the fingerlings and the fertilizer. Resourceful fish farmers could breed their own fish and use only natural fertilizer, but beginners would be more sure of success with purchased fingerlings. Annual costs are worked out in Table 10.4.

Initial construction costs will vary with the design. The drains and intake screens would probably have to be purchased. In principle everything else could be done by the owner. A conservative estimate of the total cost for the pond might be $1100 but could be less, as described in Chapter 4. Annual bank payments, including interest and repayment on the principal, could total $220 (10 years at 15%). These costs, plus fish and feeding, would bring annual costs to approximately $364:

**Table 10.4. Annual Costs/Acre for
1-Acre Pond**

|  | ($) |
|---|---|
| Stocking |  |
| Bass | 24 |
| Sunfish | 60 |
| Fertilizer | 60 |
| Annual costs/acre | 144 |

| Construction | $1000 |
|---|---|
| Supplies | $100 |
| Total | $1100 |

| Annual care costs | $144 |
|---|---|
| Bank payments | $220 |
| Total annual costs | $364 |

The cost per pound of fish from the farm is $2.43. Fish in the supermarket probably sells for $2.50 per pound. Not included in this calculation is the labor of taking care of the pond or the aesthetic pleasure of having one.

Another way to earn money from a fish pond is to allow people to fish from it for a fee. This is done in the United States and elsewhere. A pond in a large park in Dhaka, Bangladesh, contains a floating enclosure in which fish are grown until big enough for release. Along the shore of the same pond are many small fishing docks which are rented for a small fee.

# OTHER TYPES OF FISH

Bass and sunfish or catfish are not the only fish that can be grown in aquaculture. If the objective is to generate cash, alternatives to catfish are trout or salmon. Both of these command high prices but need more care than catfish and also need a more elaborate pond arrangement — cool, clean, running water. If the objective is to raise much protein efficiently, carp or tilapia can be raised. Carp are bony and not generally prized for eating in the United States but they are well regarded in the Far East and are easy to grow. They flourish in murky water, look like huge goldfish, and are sometimes grown as ornamentals.

Wild tilapia are common in East African lakes and often are suggested for aquaculture in the United States. They breed prolifically, so population control is a problem. Tilapia will not survive in water temperatures under 60°F, which means they must be wintered in a heated place, but this also means that escaped fishes are not a threat to the local ecology. Tilapia are considered very tasty and commercially raised fish are appearing on menus in upscale restaurant. Tilapia culture is described in New Alchemy reports [6].

Crayfish are raised in the South. They can be raised in rice fields because they eat rice stubble and will hide in burrows when the field is drained. Crayfish culture is described in the U.S. Fish and Wildlife report [3]. Shrimp farming has been successful in Indonesia on a small scale — the shrimp are exported to Japan [1].

Combinations of different types of fish, or fish and ducks, can be used. The intention in combining fish is to put bottom feeders with surface feeders so the pond is used efficiently and the fish do not compete. An efficient combination is carp, which lives in the middle or top of the pond, and catfish. Such a system is used in Southeast Asia. Bass flourish in clearer water than catfish, so these fish do not combine well. McLarney [5] describes a situation in which the yield of catfish in a pond increased by 12% when tilapia were added to the pond — the total yield from the pond increased by 31%.

Another aquaculture scheme used in Southeast Asia is growing fish in cages suspended in natural ponds [2]. Normal water circulation in the pond maintains the water quality. The cages keep the fish near the food supply; that is, the farmer sprinkles the fish food over the top of the submerged cage so all of it goes to the fish. The cages also, of course, aid in harvesting. The frontispiece to this chapter shows a fish cage.

## NONECONOMIC FACTORS

One reason a fish pond is appealing is its efficiency. With only a little effort and expense one can get much protein from land that might not grow anything else. Why is a pond so efficient? It collects energy from the sun better than the same area of field. Basically all food gets its energy from the sun: plants directly and animals through plants. Land plants absorb only the energy at the surface of the ground, but ponds absorb the energy through a depth of water, so more energy is trapped and available. The energy absorbed, either as heat or converted to plants, diffuses through the pond as the water mixes much more effectively than energy moves about a field. Fish convert energy more efficiently than birds or animals partly because they are cold blooded and do not use energy warming themselves. Another reason fish are efficient converters is that the water holds them up, so they do not use energy simply to stand. The efficiency argument is less compelling, however, when a pond is fed. Commercial fish food contains much energy, so the energy efficiency of an aquaculture system with supplemental feeding is lower than that of a natural pond.

Ponds are valuable for other reasons: for example, recreation, or because the owners enjoy the taste of fresh fish. They can increase the attractiveness of a piece of property. They can be part of a hydroelectric project or store water for irrigation. They can attract wildlife, such as migrating ducks and geese, or mammals, such as otters. Since some wildlife will eat fish, pond owners who like both fish and wildlife will have to make trade-offs. Fish ponds also give owners some independence from the local food store.

The environmental effect of a pond is usually positive, although one should always make major changes in streams and wetlands with sensitivity. A pond creates another habitat for wildlife and stores water. It cools the air during the summer through evaporation and absorbs and stores heat in the winter, thus moderating temperature extremes. A hazard posed by a pond, of course, is the possibility of drownings and precautions must be taken.

# REFERENCES

1. Appropriate Technology International (ATI) (1989). *Shrimp Farming in Indonesia*. ATI, Washington, DC.

   *Describes a project to promote shrimp production in brackish water ponds. The shrimps are exported.*

2. Chakroff, M. (1978). *Freshwater Fish Pond Culture and Management*, rev. ed. VITA Publications, 1815 North Lynn Street, Arlington, VA 22209.

   *Intended for development workers overseas so varieties of fish are not those described in this chapter. Readable, with helpful pictures.*

3. Dupree, H. K., and Hunter, J. V. (1984). *Third Report to the Fish Farmers*. U.S. Fish and Wildlife Service, Washington, DC.

   *Emphasis on commercial catfish raising but chapters on other fish including bass, sunfish, carp, crayfish, and even alligators; extensive bibliography; best place to start if you want to start a pond in the United States.*

4. Leckie, J., Masters, G., Whitehouse, H., and Young, L. (1981). *More Other Homes and Garbage*. Sierra Club Books, San Francisco.

   *Broad discussion of freshwater clams, buffalo fish, etc; extensive bibliography.*

5. McLarney, W. (1987). *The Freshwater Aquaculture Book: A Handbook for Small Scale Fish Culture in North America*. Hartley & Marks, Point Roberts, WA.

   *Extensive description aimed at small-scale producer; proponent of polyculture.*

6. McLarney, W. D., and Todd, J. (1977). Walton two: A complete guide to backyard fish farming. In *The Book of the New Alchemists*. The New Alchemy Institute, Box 432, Woods Hole, MA.

   *Mostly about tilapia in greenhouses; worthwhile discussion of the philosophy of raising fish in one's backyard.*

7. Report of an Ad Hoc Panel of the Board on Science and Technology for International Development (1988). *Fisheries Technology for Developing Countries*, National Academy Press, Washington, DC.

   *Mostly boats and fishing gear; a chapter on artificial reefs and one on raising fish, shellfish, and seaweed at the ocean shore.*

## PROBLEMS AND PROJECTS

1. Work out the economics of feeding rather than just fertilizing a catfish pond. Do you recommend using supplemental feeding?

2. Work out the economics of a catfish pond (i) using only fertilization and (ii) using supplemental feeding.

3. Consider a real site known to you and design a fish farm. Choose the type of fish. Decide how you will benefit from the fish farm: selling commercially, fee fishing, supplying your own home, or something else. Decide how much care you want to take. Estimate construction costs and other costs.

4. Hydroponics consists of growing plants in a fertile solution, as described in Chapter 9. Design a system combining a hydroponic system with fish raising, using fish wastes to supplement the fertilizer.

5. Plan an integrated farm using animal manure to fertilize the fish pond. How many animals are needed? How big of a pond is needed?

6. An experiment in Zambia consisted of two small ponds next to each other. During one year fish and ducks were raised in one pond and vegetables in the other. At the end of the plant-growing season the first pond was drained into the second and the fish moved. (Ducks move by themselves.) Each year the role of each pond changed. Design such a system in detail. Does it seem feasible? Does it seem worthwhile?

*Chapter*

# 11

·········

# HEALTH CARE

·····················

## INEQUALITIES IN HEALTH CARE
## THROUGHOUT THE WORLD

This chapter focuses on health problems in the Third World, particularly infant mortality. As everyone is aware, health care is not the same throughout the world. In Sierra Leone 1 in 5 children die before reaching the age of 1 and in Bangladesh it is 1 in 8 children. At the other end of the spectrum are the wealthy countries such as Japan and Scandinavia, in which 1 in 142 and 1 in 125 children die before age 1, respectively. Furthermore, in the poorest countries the infant mortality rate is about seven times the mortality rate of the overall population, whereas in the wealthy countries the rates are about the same.

Health care problems in the Third World are not the same as those in the United States. The high death rate among children in the Third World is primarily due to communicable diseases, gastrointestinal problems, parasitic diseases, and respiratory infections; these have been brought under control in the United States by technology, chiefly by antibiotics, vaccinations, and sanitation. The latter two factors are basically public health measures. Present-day Western medicine emphasizes treatment of diseases that are not the big killers in the Third World. The problem in the Third World is how to use very limited funds to treat sicknesses, e.g., malaria and cholera, which are not seen often in the United States. The question we address is whether Appropriate Technology can play a role in making health care better in the Third World.

## PRIMARY HEALTH CARE

Primary health care is the first level of contact of individuals, the family, and the community with the national health system. In a typical Third World village it is a clinic, staffed by a clinical officer — somewhat equivalent to a physician's assistant in the United States — and a midwife. The staff prescribes drugs, performs vaccinations, dresses wounds, performs minor operations, and arranges transportation to hospitals in serious cases. The staff also conduct "under-five" clinics for young children, offer prenatal and postnatal care for mothers, and does health education as practical. The clinical officer and the midwife undergo formal training, usually for 2 years. We consider primary health care because it can be consistent with Appropriate Technology — small scale, controlled by the local community, labor-intensive, and understandable by the user.

Primary health care is described in the Proceedings of a World Health Organization Conference [16].

It addresses the main health problems in the community, providing primitive, preventive, curative, and rehabilitative services accordingly. The elements of primary health care include the following:

1. Education concerning prevailing health problems and methods of preventing and controlling them, e.g., malaria, diarrhea, malnutrition, and respiratory infections

2. Promotion of food supply and proper nutrition

3. Adequate supply of safe water and basic sanitation

4. Maternal and child health care including family planning

5. Immunization against the major infectious diseases, e.g., tuberculosis, diphtheria, whooping cough, tetanus, poliomyelitis, and measles

6. Prevention and control of locally endemic diseases

7. Appropriate treatment of common diseases and injuries, plus first aid

8. Provision of essential drugs

Primary health care requires and promotes community and individual self-reliance and participation in the planning, organization, operation, and control of primary health care, making full use of local, national, and other available resources, and includes education so community members can participate.

## WOMEN AND HEALTH TECHNOLOGY

A Kenyan woman who has worked extensively in East Africa on public health problems, Float Auma Kidha [6] writes,

> *Women perform most of the tasks which have a primary and most significant impact on the standard of living. Women do the majority of work with little or no financial reward for as many as eighteen to twenty hours a day, using backward and labor-intensive technology. It follows therefore that any technology that lightens the burden of women will have a positive influence on the health of the family, community, and nation.*

## HEALTH CARE DELIVERY SYSTEM

Ms. Kidha [6] notes,

> *In many developing countries health facilities are so far away from the families that a mother would plan to be away from home for the whole day to get her child immunized or have other maternal and child health services for herself, such as antenatal care, postnatal care, family planning, etc. In many countries these services are segmented in that mothers*

*have to make multiple trips to the health facilities, for example, visit on Monday for her child's immunization, on Wednesday for Nutrition Clinic, for another malnourished child, and yet again on Friday for her antenatal care. Many times it is the same woman who must fetch water some three kilometers away, gather firewood in a forest two kilometers away, gather vegetables, go to the market, the grinding mill, etc. before preparing food and feeding the family. It is too much to expect her to walk ten kilometers to the health facility and back unless her child is very, very ill. More often than not, by the time she makes the long trip to the health facility, she has tried the village herbalist's medicine or the witch doctor's charms. The health services in rural areas still are used more for emergency curative care than for preventive purposes.*

*To raise the health standards of the family ways must be found to lighten the woman's load in one or more ways. Water, firewood, and health services must be within easy reach. Even the earthen pots for carrying water are too heavy. Women must be allowed to make decisions regarding their own fertility so that they can independently decide not only on how many children they want to have, but when they will have them. In many African cultures for example, it is the mother-in-law who decides on the number and spacing of children for the couple.*

*Health services should be arranged in such a way that all services, e.g., immunization, antenatal care, child spacing or family planning, postnatal care and curative services, etc., are all given every day. In this way, a woman can plan to go to the health facility for multiple services on a single day, which will leave her with more days at home to attend to those chores which are very necessary for the health and general living standards of her family. More days at home of itself would go a long way in preventing many deaths among children which are usually due to communicable diseases and malnutrition. An integrated system of health care delivery, especially in the rural areas of developing countries, is a technology which takes into consideration a wider perspective than health alone. It embraces the total way of life of a woman!*

*Medicinal drugs are an important aspect of medical technology — the list of drugs needed for primary health care are those drugs required in specific local circumstances. The commonest diseases for which drugs are needed in villages are malaria, diarrhea, respiratory tract infections, skin conditions, etc. Drugs for malaria, like chloroquine, as well as pain relievers like aspirins are usually kept by the local shopkeeper from whom people buy after clinic hours or when the clinic has run out of chloroquine — a situation which occurs more often than not. Because of lack of knowledge about chloroquine dosage, people buy two or so tablets which are taken in one dose only rather than the needed series. There are several reasons for single dosages. One is that people do not know the proper dosage of chloroquine, even if they had enough money. The second reason is that the shopkeeper is most interested in selling her/his stock and not advising people on dosages of the drugs she/he sells over the counter (of course, often the shopkeeper does not know the dosage). The third reason is that drug manufacturers usually write the instructions in English so that even if the illiterates were literate they may not understand. In some communities chloroquine and quinine have been used by young pregnant girls to procure abortions, thus pregnant women with malaria may not wish to take chloroquine for fear of miscarrying. The result of all these is that people do not take the right treatment for malaria. Another result of low dosages is that the parasite becomes resistant to the drugs available [currently a major problem in many Third World countries [14]].*

*The problem could be overcome by training local shopkeepers on the right dosages of the simple drugs which they sell. The shopkeepers should insist that a total course of chloroquine be purchased. The shopkeeper must know the side effects of every drug sold, to educate the customers. The drug stock should be appropriate for local health needs.*

*Traditional midwives and healers should be taught the dosages of the common drugs, and a supply of the basic drugs left with them by the local clinic. After all, more patients go to traditional medicine people, who are in every village, than to government clinics, which serve several villages. Places where the health care system has integrated traditional healers into their activities have succeeded in reducing the incidence of common conditions such as malaria and diarrhea.*

## Traditional Healers

As Ms. Kidha noted previously, many ill people in Africa seek the help of a traditional healer, possibly because they are the nearest medical help. Some traditional healers use local herbs exclusively, hence they are called herbalists. In at least some cases these herbs in fact have medicinal properties recognized by Western physicians, although dosages are uncertain and the drugs are not standardized. A herbalist learns the trade by apprenticing to an established herbalist, often a parent. A traditional health facility is shown in Fig. 11.1 ("Chitapala" means "clinic"). Some herbalists

**Fig. 11.1. Traditional health facility.**

claim to use supernatural forces as well as herbs. These herbalists are especially effective in places where people believe illnesses can be produced by an enemy putting a spell on a person.

It is difficult to determine the effectiveness of herbalists. They tend to guard their knowledge. Seldom do they keep careful records about treatments and results. On the other hand, they usually offer comfort and nonthreatening cures. Some African people prefer to use traditional healers to treat one type of symptom and Western doctors to treat another type: The less specific the symptom, the more traditional healers are desired.

Traditional birth attendants are also used widely in Africa and throughout the Third World. They also usually learn by apprenticing themselves to an experienced person and also prescribe herbs when needed. Traditional birth attendants are often especially effective because they are well accepted by local people.

Modern medical organizations in some Third World countries often simply ignore traditional healers or birth attendants, except perhaps when they research the pharmacology of the plants used [1, 13]. The Ministry of Health in Malawi has, however, held meetings for traditional healers. At these meetings an attempt is made to have the healers share some of their knowledge with one another. Also at these meetings, common diseases and their treatments are described. Training of traditional birth attendants is becoming more common in many parts of the Third World.

Third World hospitals sometimes exemplify a low-technology approach in another way—use of mothers and the entire family to care for sick children. In many hospitals the expectation is that the mother will feed, cleanse, and generally provide for a sick child. The hospital provides rudimentary living accommodations for the mother and siblings of the patient. Families tend to stay with sick children during the entire day. Patients, even premature infants, receive much attention and comforting. One reason this approach is used is the shortage of hospital staff, but another reason is the benefit to the patient. Studies in the United States indicate that premature babies that are stroked or handled gain weight 21.5 percent faster than those which are left alone in an incubator.

## EDUCATION

A strong link exists between the incidence of health problems, as reflected in infant mortality rates, and education. One demonstration of this link is the United Republic of Tanzania, which has one of the highest rates of enrollment in secondary schools in the world. Despite being one of the poorest African nations (per capita annual income of $240) it has the third lowest infant mortality rate in sub-Saharan Africa. Apparently education makes parents more aware of the consequences of illnesses, the causes, and the availability of health facilities. Schools promote a concern for hygiene, nutrition, and disease prevention.

Education is the only approach in some health-related situations. AIDS, of course, is a disease which is deadly and for which no cure is known. People must be educated, within the context of their own cultures, on the best way to prevent its spread and to protect vulnerable groups. Incidentally, health care workers in the Third World are concerned that the panic induced by AIDS will divert all resources to it, neglecting other health problems.

For diseases other than AIDS, what can be done? An obvious solution is to introduce Western-style medicine but the situation in the Third World (e.g., finances, water sources, and nutrition) is different from that in the Western world. Successes in improving health in the Third World will come through public health measures, for which education is an absolutely essential component. Expensive medical facilities, such as intensive care units, will not accomplish nearly as much as will knowledge by the populace of causes of diseases and elimination of these causes. In the remainder of this chapter, we consider diarrhea treatment, infant care, sanitation, and water supply. Education is an essential component of the success of any of these approaches.

# INFANT DIARRHEA

In rural villages in Africa, infant mortality is high; perhaps 150 of 1000 babies die before reaching the age of 1. Many of these children die from diarrhea dehydration. What can be done? A closer look at a village may help us realize what can be useful. Most of the people are subsistence farmers with a diet restricted to a few vegetables and grains and a little meat and fish. Their resistance to illness is low. The lack of running water in homes makes cleaning difficult, although in some African societies each meal begins by passing a bowl of clean water for rinsing hands.

Three beliefs prevalent in village society are relevant to infant diarrhea. The first is that people do not realize human wastes carry disease. The general awareness of the cause of disease may be low, and the concept of germs may not be known at all. The second belief, present in many cultures, is that diarrhea is best treated by withholding fluid intake. Such treatment, of course, increases the severity of dehydration. The third belief, that commercial infant formula is more effective or convenient than breast-feeding, has created a trend away from breast-feeding.

Besides diet, lack of running water, and the beliefs previously discussed, the proximity of a health center to the village is an important determinant of mortality from infant diarrhea. Infant diarrhea can kill quickly, so timely treatment is important. As previously noted, health centers can be a long walk from homes. Mothers take care of infants and also work in the fields and do household chores. Thus, it may take some time to realize that a baby is very sick and to get her or him to the health center.

The result of all these factors is that dehydration due to acute diarrhea is a significant factor in infant mortality. It is the most prevalent childhood disease. In 1980, 5 million children worldwide — approximately 10 children per minute — under the age of 5 died from acute diarrheal dehydration. In many less industrialized countries diarrhea disease is one of the two most common reasons for visiting health clinics, pharmacies, and hospitals.

## *Diarrheal Dehydration*

We now consider diarrheal dehydration in more detail. Acute diarrhea is an attack of sudden onset which lasts usually 3–7 days, but may last up to 14 days. Death is usually caused by two factors: dehydration and malnutrition. Dehydration is directly linked to the fluid loss as well as to the common practice of withholding fluids from the patient. Malnutrition is due to the victim's loss of appetite and the body's inability to absorb food properly.

Diarrhea is caused by approximately 25 different pathogenic parasites, virus and bacterial, which cause diarrhea by various means — usually by damaging the lining of the walls of the intestine, thereby causing the malabsorption of food and thus diarrhea and vomiting. These pathogenic organisms are transmitted in many ways. The most common methods are either by contamination of drinking water or through contaminated foods. Food contamination results from flies moving from animal or human feces to the food.

The most likely source of contamination in a village is an open latrine, from which disease can spread in different ways. Improperly built latrines may contaminate groundwater supplies, i.e., shallow wells, if the pit is not properly lined with concrete or some other appropriate material. Open latrines give flies access to the feces and the flies can land on uncovered food. Another source of contamination is feces from livestock roaming around the village.

The treatment for diarrhea dehydration in U.S. hospitals includes drugs and intravenous (iv) therapy. In most of Africa antidiarrhea drugs have drawbacks. They are imported and therefore expensive and hard to obtain. They are often ineffective against local diseases. Intravenous therapy is very effective but requires a skilled therapist as well as usually imported medical supplies such as sterile needles, iv bottles, rehydration fluid, and so forth. Intravenous therapy is also time consuming.

## Oral Rehydration Therapy

An alternative treatment for diarrhea is Oral Rehydration Therapy (ORT), which basically consists of giving the victim a large amount of water mixed with salt and sugar [2]. Parents and children can make the mixture very easily. People involved can understand what has to be done and can act rather than watching the victim die.

Oral rehydration therapy, although simple and inexpensive, is effective. Most patients (90–95%) can be treated with ORT alone, regardless of the cause of the diarrhea or the age of the patient. The average cost of treating one patient with ORT is (50¢) whereas intravenous therapy costs approximately $5.00. In areas where ORT programs have been implemented, there have been substantial decreases in infant mortality.

The principle of ORT consists of the replacement of fluid and essential salts that are lost during acute diarrhea. A dehydrated patient is treated in two phases. *The rehydration phase* consists of the replacement of fluid and the essential salts lost by vomitus and stool production. The *maintenance phase* follows, in which one compensates for continuing abnormal losses due to diarrhea and vomiting as well as losses due to normal

If you lift up the skin and you can still see the fold after you let go, the child is dehydrated.

## How to prevent dehydration in diarrhoea

Many of the herbal teas and soups that mothers give to children with diarrhoea do a lot of good because they help get water back into the child. Breastfeeding provides both water and food and should always be continued.

KEEP BREAST FEEDING

A BABY WITH DIARRHOEA

Demonstrations help children understand how important it is to give a child with diarrhoea as much water as he is losing. For example:

As long as just as much water is put back as that which is lost, the water level will not go down (so the child will not get dehydrated.)

A child with diarrhoea needs 1 glass of liquid for each time he has a loose stool.

Giving lots of liquid to a child with diarrhoea may at first increase the amount of diarrhoea. This is all right. *THE DIRTY WATER MUST COME OUT.* The important thing is to *be sure that the child drinks as much liquid as he loses*.

## The Special Drink
### How to mix it

The Special Drink, made from SUGAR, SALT and WATER is especially good for children (or adults) with diarrhoea.
It is simple to make:

MIX:  SUGAR    +    SALT    +    WATER

one level teaspoonful of sugar  +  a little at the end of a spoon  +  one glass of water

OR  a scoop of sugar  +  a pinch of salt  +  one glass of water

*A special plastic spoon will be available. Use this to teach everyone the right amount of sugar and salt for each glass or cup.*

*BEFORE GIVING THE DRINK TASTE IT - IT SHOULD BE NO MORE SALTY THAN TEARS*

## How to give the Special Drink

Start giving the Special Drink AS SOON AS diarrhoea begins. A child should drink one glass for each stool he passes.

If the child vomits the drink, keep giving him more. A little of it will stay in his stomach. Give it in sips every 2 or 3 minutes. If the child does not want to drink, gently insist or coax him to do so.

CHILD                        ADULT

One glass each stool        Two glasses each stool

Keep giving the drink every 2 or 3 minutes, day and night, until the child urinates normally (every 2 or 3 hours.) Older children and their mother can take turns through the night.

## Warning signs

Take child with diarrhoea to the health centre if he:
1  SHOWS ANY OF THE SIGNS OF DEHYDRATION
2  Cannot or will not drink
3  Vomits so much he cannot drink
4  Makes no urine for 6 hours (time from dawn to noon, or noon to dark)
5  Has diarrhoea so often he cannot drink one glass per stool
6  Has blood in his stool
7  Diarrhoea lasts more than 2 days

## Figuring out how well the children are doing

Encourage the children to discuss how they can find out such things as:
• how much they have learned;
• whether other people in the community have learned some of the same information;
• how many of the children have put their new knowledge (about diarrhoea) to work;
• if fewer babies and children suffer and die from diarrhoea as a result of this activity.

If this discussion takes place when they are beginning the activity they will be able to gather the information they need to make comparisons as they go along.

Counts can be made each month (or after six months or a year) to see, for example:
• how many children (or their mothers) have made the 'special drink' for those with diarrhoea?
• how many cases of diarrhoea have there been?
• how many children have died?

Ask any child who has used the Special Drink for a brother or sister with diarrhoea to tell the story to the school, explaining how he (or his mother) made and used it and if it seemed to help.

*Ref: AS3E*        *CHILD-to-child January 1979*

## Fig. 11.2. Instructions for preparing ORT mixture.

activity—sweating, breathing, and urination. It is important to continue giving the mixture until the diarrhea passes. One guide is to continue giving until the patient urinates normally, every 2 or 3 hours.

Of course, ORT is not a cure for acute diarrhea. Diarrhea only lasts for a limited time and the ORT helps the patient live through the effects of the illness, not cure it.

## The ORT Mixture

The ORT mixture is simply salt, sugar, and water and thus can be prepared at home. A page from a manual for health workers in the Third World is shown in Fig. 11.2. It shows how the mixture is made and administered. It also suggests how a health care worker can explain the treatment. Such pictorial explanations are, of course, important, especially if the village people cannot read. Another mixture is described at the end of the chapter in Fig. 11.9 and Problem 5. An advantage of having those who need the mixture prepare it is that powerful members of the village cannot withhold it from those less powerful.

## Packets of ORT Mixture

Some health care workers are not in favor of expecting people to prepare their own ORT mixture and advocate the use of prepackaged single doses. A packet containing such a dose is shown in Fig. 11.3. This packet actually came from a health center in a camp serving refugees from Mozambique.

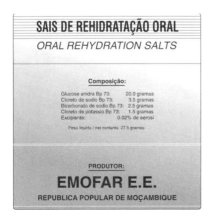
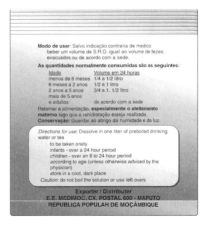

**Fig. 11.3. ORT single-dose packet.**

Ms. Kidha, who wrote about women and health care earlier in this chapter, believe these packets make the best sense. She writes [6];

> *Packets of oral rehydration salts can be left with the local shopkeepers, traditional medicine man or traditional birth attendant after they have been trained in its use. These packets would be available within reach should a child start having diarrhea. Homemade sugar–salt solutions are not a practical way of telling a village woman to deal with diarrhea. Sugar is not found in many households in some villages. Salt is a necessity and might not be "wasted" for dehydration. Most importantly, it is very difficult to find a uniform container to measure one liter of water in the villages. This leaves room for errors which may prove fatal. For example, a mother brought her child unconscious into a Pediatric Ward after giving it a sugar–salt solution she had made at home using 10 teaspoonfuls of sugar, one teaspoonful of salt and instead of adding water from three full coke bottles, she added only one! The mixture was not the right concentration for the baby so ....!*
>
> *It is expecting too much of a busy mother who has to worry about not only her child with diarrhea but a whole household to remember at that anxious time to mix ten sugars, one salt with one liter of "boiled" water! In the first place firewood is scarce and it would take quite a time to get the fire going at three in the morning. Then wait for the boiled water to cool before it is given! May be too late! Mothers should be taught to give any drink available in case of diarrhea, like thin porridge, black tea, rice porridge, etc. and not waste time boiling water or looking for sugar etc. Every home has some fluids which will prevent dehydration at a time like this.*

We have discussed how to treat diarrhea. Now we consider how to prevent it. Three approaches seem effective: breast-feeding, pit latrines, and a clean water supply.

## BREAST-FEEDING

Mothers normally have a choice between breast-feeding and the use of a commercial infant formula [15]. The benefits of breast-feeding are straight-forward. It is much healthier for the baby because the milk provides all the essential nutrients and is not contaminated, whereas the water used in preparing formula may be contaminated. Infant formula is expensive, approximately $200–300 per year, which is perhaps 30% of the monthly salary for an average government worker in an African country. Breast-feeding protects the child against other diseases besides diarrhea since the mother's natural immunity (antibodies) is passed to the infant through the milk. Finally, breast-feeding helps prevent pregnancy. Women who do not breast-feed have a postnatal infertility period of 3 months, whereas women who breast-feed are usually postnatally infertile for a period of 13 months, but this usual period of infertility is not reliable enough to be depended on.

The reason that mothers do not breast-feed as much as they could seems to stem from well-intentioned, but misguided, aid programs in the 1960s. Well-regarded medical research indicates that a 6-month-old baby needs more protein than a mother's milk can provide, so some kind of supplement is helpful until the infant is entirely weaned at approximately 24 months. Locally made supplements of cow's milk, thin rice gruel, or the best commercial formulas are good supplements to mother's milk for an older infant. Formula actually contains slightly more nutrients than mother's milk but no antibodies. In any case, formula is not needed before the baby is 6 months old, in normal situations, and formula was originally intended as a supplement.

A problem with formula is that clean water and sterile bottles and nipples are hard to obtain in many villages. Furthermore, once the formula is made, it will spoil unless kept cool and refrigeration is often not available. Also the correct proportion of mix to water must be used but not all parents are able to read the mixing instructions.

Multinational companies are sometimes accused of having jumped into the formula market, opened by the aid organizations, with aggressive and possibly misleading advertisements of the necessity of formula. Certainly multinational companies have been severely criticized for promoting formula in the Third World [8]. Just as in the United States, some parents in the Third World perceive anything modern and imported as better. Formula does tend to be more convenient because feedings are required less often on a routine schedule. So, for a variety of reasons, some mothers favor formula over mother's milk. The need exists to educate people about the relative advantages of both mother's milk and formula. Recent reports seem to indicate that breast-feeding is now being done more often than in the recent past.

## THE JOYS OF OWNING YOUR VERY OWN PIT LATRINE

The proper disposal of human and animal wastes is a major sanitation problem in the Third World and elsewhere [4]. Methane digesters, described in Chapter 8, are an approach to the animal waste problem but are less effective in killing the pathogens of human wastes, which are a larger health problem. Pathogens from human wastes, if they get into the water supply, can cause cholera and hepatitis as well as diarrhea. Flies can also spread disease, by carrying germs when flying from feces to food.

Most people in the world probably want flush toilets emptying into septic tanks or sewers, but the cost of providing such systems to everyone in the world makes such toilets out of the question. A World Bank report [11] estimates that 2 billion people worldwide need improved sanitation facilities. The cost of providing sewers would be about $500 per person, and the annual per capita income of half of these people is less than $200. Septic tanks would not be appreciably cheaper. The ventilated improved pit latrine (VIP) described next and in Ref. [11], developed in Zimbabwe for rural use, costs about $100. Even a more substantial version useful in, for example, a settlement on the edge of a city, can be built for $150.

## Design of Ventilated Improved Pit Latrine

A conventional pit latrine consists of a pit, a squatting plate, and a superstructure. Problems with a conventional pit latrine are odors and insects. The ventilated improved pit latrine shown in Fig. 11.4 solves these problems. The vent pipe shown is the key component. The vent pipe, which is painted black, is heated by the sun so the air rises. Wind blowing across the top of the vent also pulls air out of the pipe. Air is thus pulled down the squat hole through the pit and out the vent. Inside the superstructure no odor exists because of the direction the air is moving. A way of augmenting this airflow is to locate the door opening so the prevailing wind will be caught to blow air into the pit.

The vent pipe is screened at the top. This prevents insects from escaping and also keeps them from coming down the vent. Some flies will get into the pit through the squat hole and lay eggs. The emerging flies, though, will be attracted to the light, and will fly up the vent pipe and be caught at the screen. The other important insects are mosquitoes. If the pit is dry, mosquitoes are not attracted. If the pit is wet, that is, if the level of groundwater is above the bottom of the pit, mosquitoes will breed there. Mosquitoes are not attracted as strongly to light as are flies, so some will go out the hole. In this case the hole should be covered with a removable screen. It is important, of course, not to interfere with the circulation of air through the system by using a solid hole cover.

A necessity is relative darkness inside the superstructure, so the flies are attracted to the vent pipe. If social custom favors an illuminated superstructure, then some sort of opaque cover will be needed over the

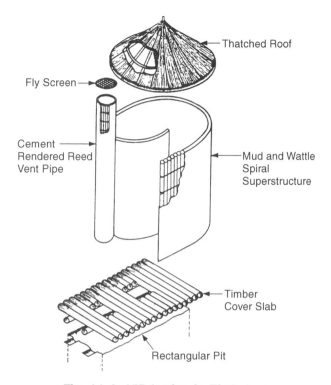

**Fig. 11.4. VIP latrine in Zimbabwe.**

hole. This cover should be raised from the floor to allow as much ventilation as possible using screening to keep insects in the pit. If the entrance to the latrine faces east or west, the morning or evening sun may shine into the latrine, so shading may be necessary. If wind conditions permit, north or south orientation should be used.

Lighting is an important issue. When VIP latrines were built near a health clinic in Kenya, care was taken to make them very dark inside. Snakes then found the superstructures desirable places to live. The local people in turn refused to use the latrines. One lesson is that a design must meet local conditions.

If the pit is above the groundwater level, the pit will be dry and the wastes will decompose in the pit. If the pit is below groundwater level, the pit will have water in it and some of the wastes will seep away into the ground, just as they do in septic systems not connected to sewers. In this case, wells for drinking water should be at least 150 ft from the latrine.

A dry pit 10 ft deep and 3 ft long in either direction should last for about 10 years for a family of six. The useful life will be twice as long if the pit is wet. (In designing, one should allow for about 1.5 cubic feet of waste per person per year.) The VIP latrine is so simple to build that one might build a new one somewhere else when the pit is full. Alternatively, one could build a double-pit latrine and use each pit alternately for a year. At the end of a year, the material in the unused pit will have decomposed and could safely be removed and used as fertilizer, if social customs permit.

## Construction of the Rural VIP Latrine

The basic rule to follow is to make the construction as similar as possible to other structures in the village, so the model shown in Fig. 11.4 would be used when the surrounding buildings are made of mud and wattle. The vent pipe is made of a reed mat, 8 × 3 ft, rolled up into a cylinder about 11 in. in diameter, and then plastered with a mortar made with cement. Ordinary corrosion-resistant screening can be used at the top. The rest of construction is obvious. The roof thatching has to be dense to keep the interior dark. Of course, people who use thatching regularly will have no trouble constructing this. The walls should not have light leaks. The people building the structure will have had experience in making tight walls. If the soil crumbles easily the top 3 ft or so of the pit should be lined with cement. The logs supporting the timber cover slab should extend approximately 1 ft in each direction beyond the pit hole. They should be treated with a wood preservative. The slab itself should also be painted to protect the wood.

In areas where wood is scarce, the superstructure can be built of thatching or of locally made bricks. In more urban areas, cement over a wire frame can be used, and the squat plate can also be made of cement. A latrine with a cement squat plate in a refugee camp is shown in Fig. 11.5. In the figure, the latrine is under construction and the squatting holes are covered with mud or bricks. The vent pipe can be asbestos cement or PVC; cast iron corrodes. Table 11.1 shows the costs of the VIP latrine. It would take three persons about 1 week to build the latrine.

## Success of VIP Latrines

These latrines have worked well in Zimbabwe, where 20,000 were in use in 1982. Perhaps one reason for their success is that they represent a relatively small change from what had previously been in use. Another reason is probably the extensive education program that the government of Zimbabwe used, including films, instruction leaflets, and demonstration models.

**Fig. 11.5. Latrines under construction.**

**Table 11.1. Cost of VIP Latrine**

| Material | Cost ($) |
|---|---|
| Cement for pit lining and vent pipe (55 lbs) | 2.00 |
| Cement for superstructure (55 lbs) | 2.00 |
| Wire (150 ft) | 0.70 |
| Fly screen (1 × 1 ft) | 0.20 |
| Nails | 1.00 |
| Paint | 1.00 |
| Total | 6.90 |

Disposal of human waste is a more complex cultural issue than many aid workers realize. Ms. Kidha [6] writes about her experience,

> *There are customs and beliefs about human waste disposal and even where people know the connection between human waste and diseases, they have difficulty overcoming their beliefs. Health workers have failed to reduce diarrheal diseases by forcing people to build latrines. People have built latrines for fear of the authorities but not because they understand why*

*they must build them. The result is that nobody used them. Health education is the key. But the health educator must know and appreciate the culture of the people and educate them in the context of that culture. In many cultures it is believed that using a pit latrine is like being buried alive. They believe that human waste must go back to the land and fertilize it. For this reason the bush is used more often. People can be taught to bury their waste if they will not use pit latrines. Another belief is that a man would get ill and even die if he used the same latrine which has been used by a woman who is menstruating. A young woman would never use the same latrine which her mother and father-in-law use. Human waste is used for manure in many cultures. It is, therefore, not fact to assume that one's own pit latrine will ensure reduction in diseases due to poor sanitation.*

*People built latrines but never use them for various reasons. In one village in Kenya, health workers forced people to build latrines during a cholera outbreak. When the health workers returned to the village two weeks later every one had a latrine which the health workers duly noted. It turned out later that the round "huts" had holes which were only three feet deep and were not being used as latrines but stores for illegal brew! People did not believe latrines were useful for health purposes but they did need a place to store the homemade distilled beer.*

Education and recognition of local customs are essential for success.

## WATER SUPPLY

Safe water improves health. Villagers usually collect water from whatever water source is within reach. More often than not the same water source is used for watering animals, bathing, swimming, washing clothes, and, of course, for cooking and drinking.

Successful water supply projects have usually consisted of wells and pumps, although in mountainous areas, clean water from streams can be brought down the mountain to villages in pipes [10]. Figure 11.6 shows a village pump. Locally made pumps were discussed in Chapter 6.

Ms. Kidha [6] notes,

*In Third World villages, usually water is not boiled for drinking. In fact many people do not like the taste of boiled water for drinking. Another reason for not boiling water is fuel. When there is so little firewood to cook*

**Fig. 11.6. Village pump in Malawi.**

*for so many family members it is impractical to ask mothers to boil water. So water is drunk the way it is gotten from ponds or rivers, polluted in many instances. When water is obtained from wells, it is often clean so normally very little is done to it in the home before it is drunk. The problem is that when the pump has broken down the villagers revert to any source of water, whether it is safe or not, because they have become accustomed to a safe source. People seem to build up a tolerance to some pathogens in the water so reverting to a contaminated source is more harmful than continuing to use one.*

Visitors to Third World villages are advised to drink bottled spring water, soft drinks, or tea, if firewood is available.

## *Water Purification*

It is sometimes necessary to purify water. Two common ways to do this are to use chemical disinfectants and filter beds. Chlorine compounds are common disinfectants but are not generally available in the Third World.

Boiling, of course, kills germs, so a combination of boiling and passing through a filter is an effective means of purification but not often used in villages. Filters alone make the water look clean but do not remove germs. Layers of clean sand make an effective filter. Charcoal mixed with the sand will also remove odors. Purification is discussed in the *Village Technology Handbook* [17].

Water is purified by distillation. If the water is very poor quality, for example, from stagnant pools or from the ocean, distillation may be the only practical way to make it drinkable. Small solar stills are fairly easy to make (Fig. 11.7). The sun's energy goes through the glass windows and evaporates the impure water in the pool at the bottom of the still. The evaporated water condenses on the glass windows and flows along them into the channels at the bottom, where it is collected. As long as the glass is tilted at least 20° the water will flow along it, rather than drip back into the pool of impure water. A reasonable size still will supply sufficient water for one adult, so they are useful in emergencies but probably are not practical for a village water supply [9]. In the Kalahari desert village well water was much too salty to be drunk. A solar distillation facility was established which earned cash by selling the salts remaining after distillation.

Another low-technology way to purify water is to use plants [12]. Water hyacinths, for example, extract impurities from the water (Fig. 11.8). The hyacinths then can be harvested and composted or fed to a methane digester. On the other hand, one must be very careful about introducing water hyacinths, or exotic plants in general, because they can multiply rapidly and clog streams and ponds. This escape has taken place in Florida, where water hyacinths have become a major environmental problem.

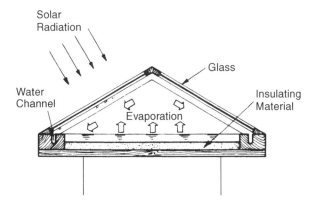

**Fig. 11.7. Solar still.**

**Fig. 11.8. Water hyacinths in Bangladesh.**

## *Rainwater Collection*

Another approach to supplying clean water is to collect rainwater, usually from a roof. Systems for doing this are used in Thailand, New Zealand, and other places. The major design issue is determining the size and construction of the storage tank. In Thailand, for example, large baskets are coated with cement, resulting in tanks of 1000 gallons capacity [3, 4].

During the first few minutes of a rain storm, the water from a roof, will rinse dirt off the roof and will be dirty. It must be determined how to divert that water from the storage tank. One device uses a 1-gallon bucket to take the first flow; when it is full it tips over, directing the rest of the flow to the storage tank [3].

## *Design of a Village Water Supply*

Designing a village water supply is straightforward but somewhat complicated because so many choices are possible (Table 11.2) [7]. One must

**Table 11.2. Design Decisions for Rural Water System**

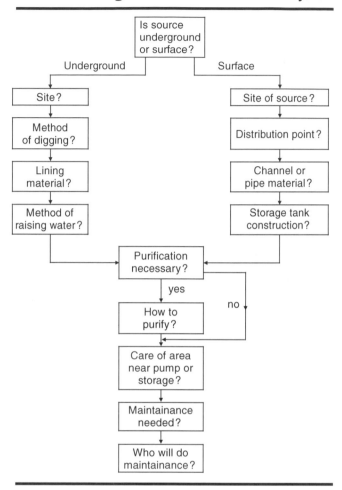

choose a source, either surface water from a pond or stream or underground water brought up by a pump or buckets. The hardest part of building a well may be determining a good location. Hiring a hydraulic engineer or a geologist may be the best tactic. The VITA book [17] gives the following advice: Try to find a place where water will collect underground, near rivers but above bedrock or above layers of impervious clay.

If underground water is selected then one must decide how to get to the water—drilling or digging. A drilled well is about 4 in. in diameter, whereas a dug one at least 3 ft in diameter. A well can be drilled by hand or machine. A tube or pipe is used as a liner for drilled wells, partly to prevent cave-in but mostly to keep contaminated surface water out. These are called tube wells. The tube or lining is usually PVC plastic, although metal can be used.

Wells can be dug by hand. An open well is easily contaminated by surface water or other things falling in. Open wells can be hazardous to young children and animals. Dug wells, therefore, are often sealed at the top. Again, they usually must be lined. An advantage of a dug well is its larger diameter, which means more water is stored than in a tube well. Storage is useful if the water seeps into the well slowly. A disadvantage is the expense of the large lining, usually made from brick or concrete. A deep well will reach cleaner water and a well as deep as 200 ft can be dug, but a maximum depth of 50 ft is more common. If one person is doing the digging, the diameter can be 3 ft but if two people are digging the diameter should be approximately 6 ft.

A pump is the most common way of rasing water (see Chapter 6). Water could be raised from an open dug well by a bucket but the risk of contamination of the open well is high. In any case, some way of raising the water must be chosen.

If surface water is chosen, the water source can be a nearby pond or stream. It can also be a mountain stream some distance away. In Chapter 13 a water supply scheme in Malawi [10] is described which consisted of laying PVC pipes from streams high in the mountains down to the villages. The major problem with surface water, of course, is its purity. Running water can be clean if sources of pollutants upstream are far enough away, but one should test the water carefully before drinking it. Pollution may be less of a problem in the Third World, where fewer factories discharge chemicals into streams and less fertilizers, herbicides, and pesticides are put on fields.

The design decisions in a surface water system, besides the water source, are how to build the channel taking water from the source to the village and what kind of water storage tank to build in the village. The location of the storage tank, that is, the distribution point, must also be selected with the advice of the users.

Other design questions exist. Is purification necessary? If so, how should it be done? Another cleanliness issue is the area near the water source; how does one avoid mud or stagnant pools which breeds insects? The pump in Fig. 11.6 is set on a large concrete pad to avoid muddy ground around the pump. Finally, one must decide what maintenance will be needed and who will do it.

A review of successful water projects emphasizes the need to involve the people who will use the system. Not only is their labor needed but also their advice regarding suitable locations, both for sources and for the distribution points is needed. Advice on how water is used and on how best to organize maintenance is essential.

# COMMUNITY PARTICIPATION

In all health-related projects, not just water supply, it is necessary to involve the future users. Ms. Kidha [6] has also had experience in this area and notes,

*Communities vary greatly in size and their socioeconomic profile range from clusters of isolated homesteads to more organized village, towns, and city districts. Community participation in deciding on policies and in planning, implementing, and controlling development programs is now a widely accepted practice. The effectiveness of such participation is greatly influenced by the overall political structure and the socioeconomic situation of a country.*

*Community participation is the process by which individuals and families assume responsibility for their own health and welfare and for those of the community and develop the capacity to contribute to their community's development.* They come to know their own situation better *and are motivated to solve their own development instead of being passive beneficiaries of development aid. They need to realize that they do not have to accept conventional solutions and technologies that are not appropriate but can improvise and innovate to find suitable solutions. The community must be able to appraise a situation, look at the various alternatives, and estimate what their own contributions can be. The government health system, on the other hand, is responsible for explaining and advising or providing information about favorable and adverse consequences of proposed interventions as well as their relative cost.*

*There are many ways in which the community can participate in their care. It must first be involved in the assessment of the situation, the definition of the problem, and the setting of priorities. It then helps to plan health care activities and finally cooperates when these activities are carried out. Such cooperation can contribute labor as well as financial and other resources. It also includes the acceptance by individuals of a high degree of responsibility for their own health care, e.g., by adopting a healthy lifestyle, making use of health services, etc.*

It is therefore important, as mentioned earlier, that the "technology" must be acceptable to those who apply it and to those for whom it is used—the community. The most productive approach for ensuring that Appropriate Technology is used to start with the problem—identified by the people — and then seek and develop a technology which is acceptable to the community and can be used by its members.

# REFERENCES

1. Gelfand, M., *et al.* (1993). *The Traditional Medical Practitioner in Zimbabwe: His Principles of Practice and Pharmacopoeia.* Mambo Press, Gureru.

   *The first part is a discussion of what traditional doctors do. The second part is a survey of plants used as remedies; a great deal of information is included.*

2. Hirschhorn, N., and Greenough, W. B., III (1991, May). Progress in oral rehydration therapy. *Sci. Am.* **264**(5), 50–56.

   *Various alternatives; describes the chemistry in the intestine; solid science.*

3. Institute for Rural Water (1982). Constructing, operating, and maintaining roof catchments, Water for the World Technical Note No. RWS.1.C.4 USAID. Request from Development Information Center, Agency for International Development, Washington, DC 20523.

   *Many of clear pictures; a good place to start when planning a water supply based on collecting rainwater.*

4. Kalbermatten, J. M., Julius, D.-A. S., and Gunnerson, C. G. (1980). *Appropriate Technology for Water Supply and Sanitation, a Sanitation Field Manual.* The World Bank, Washington, DC 20433.

   *A summary of toilet technology, with drawings plus good advice on choosing a technology. Clear and sensible, accessible to a novice, useful to both an engineer and a policy maker.*

5. Keller, K. (1982). Rainwater Harvesting for Domestic Water Supplies in Developing Countries, WASH Working Paper No. 20. Water and Sanitation for Health Project, Arlington, VA.

   *This report includes an annotated bibliography; several articles on large tanks, not only in Thailand but also in Arizona and southern Africa.*

6. Float Auma Kidha, personal communication, March 1989.

   *Ms. Kidha is Program Officer for AIDS at the International Planned Parenthood Foundation, Box 30234, Nairobi, Kenya.*

7. Idelovitch, E., and Ringskog, K. (1997). *Wastewater treatment in Latin America: Old and new options.* The World Bank, Washington DC.

   *Intended primarily for mangers and policy makers, with a review of technology options.*

8. Interagency Group on Breastfeeding Monitoring (IGBM) (1997). *Cracking the Code, Monitoring the International Code of Marketing of Breast-Milk Substitutes.* IGBM, London.

   *Examines how breast-milk substitutes are marketed in Bangladesh, Poland, South Africa, and Thailand. Some companies are violating the code.*

9.  Leckie, J., Masters, G., Whitehouse, H., and Young, L. (1981). *More Other Homes and Garbage*. Sierra Club Books, San Francisco.

    *Filtration and Chemical disinfection are described on pages 318–319. The focus of the chapter on water is constructing one's own water supply in the United States. Solar stills are described on pages 303–307 with diagrams; thorough coverage.*

10. Liebenow, J. G. (1981). Malawi: Clean water for the rural poor. *AUFS Reports* **40** Universities Field Staff International, PO Box 150, Hanover, NH 03755.

    *Focuses on the organization, but the technology can be inferred.*

11. Morgan, P. R., and Mara, D. D. (1982). *Ventilated Improved Pit Latrines: Recent Developments in Zimbabwe*. Technology Advisory Group, The World Bank, Washington, DC.

    *Provides working drawings, pictures, and advice; specialized but highly useful for someone interested.*

12. National Research Council (1976). Making aquatic weeds useful: Some perspectives for developing countries, Report No. PB 265 161. Prepared for the Agency for International Development, Washington, DC. Order from National Technical Information Service, Springfield, VA, 22161.

    *Useful introduction; besides wastewater treatment, considers weeds used in aquaculture and in methane digestors.*

13. Osuhor, P. C., and Osuhor, A. (1982, April). In contact with traditional medicine. *Tropical Doctor*, 81–84.

    *Describes the plants used by herbalists in Nigeria and what the herbalists do.*

14. Shell, E. R. (1997, August). Resurgence of a deadly disease. *Atlantic Monthly* **280**(2), 45–60.

    *Malaria is coming back in the Third World and coming to America.*

15. UNICEF (1983). *Assignment Children; A Journal Concerned with Children, Women and Youth in Development*. UNICEF, Villa Le Bocage, Palais des Nations, 1211 Geneve 10, Suisse.

    *This issue is concerned with ORT, breast-feeding, and immunization. The articles are aimed primarily at public health workers in developing countries. Well done and contains references.*

16. UNICEF/WHO (1978, February). Report of the International Conference on Primary Health Care at Alma Ata , USSR, September 6–12. Reproduced in "UNICEF/WHO joint study on primary health care," (E/ICEF/L, 1387) UNICEF.

17. Volunteers in Technical Assistance (VITA) (1981). *Village Technology Handbook*. VITA, 3706 Rhode Island Avenue, Mount Rainer, MD 20822.

    *The book was written by volunteers in various Third World countries. Much of the water supply chapter seems focused on Southeast Asia. The chapter describes choosing a site for a well, drilling or digging the well, lifting the water, storing it, and purifying it.*

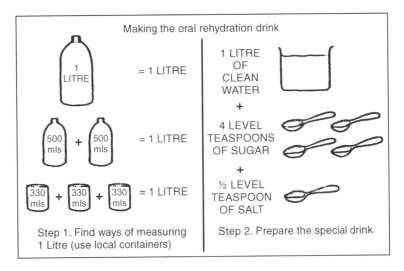

**Fig. 11.9. Illustration of another form of ORT.**

# PROBLEMS AND PROJECTS

1. Compare the infant mortality rates in different sections of the United States and make conjectures about the causes of these differences.

2. Explain in detail why health care should be approached differently in the Third World and the United States.

3. Discuss the advantages and disadvantages of training a large number of clinical officers and midwives rather than a smaller number of physicians in the Third World.

4. Develop a strategy for teaching people in a rural village to use ORT.

5. Figure 11.9 shows the preparation of another form of ORT. Compare the advantages of this form with the prepackaged mixture and the "pinch of salt and scoop of sugar" form in terms of costs and convenience to the user, risk of contamination or of improper proportions, local control, need for training, and so forth.

6. Consider why mothers may prefer to use infant formula. What steps could a public health worker take so that these preference are met when breast-feeding is used, for example, a private area at a place of employment?

7. Develop a village-level campaign to promote breast-feeding.

8. Design a VIP latrine for use in the outskirts of a city where wood is scarce. Bricks and cement are available, as is metal sheeting for roof. Estimate the cost of such a latrine using U.S. prices.

9. Design a VIP latrine system for a rural school with 50 students and 6 teachers. This is a boarding school so everyone is a resident.

10. Based on rainfall data for your area, estimate if collection of rain from the roof of a typical house would give sufficient water for the inhabitants of that house. How large a tank would be necessary? Assume each person in the house uses 25 gallons a day.

11. Prepare a detailed plan for a hydraulic engineer wishing to build a water supply system for a rural village. What should he/she do first? second? and so on. The engineer is knowledgeable about water resource problems but knows nothing about the village.

12. Design a wastewater treatment facility, using hyacinth plants, for a community of 20,000 people. Explain what you will do with mature plants that have died.

13. The recommendations of Ms. Kidha about community participation in health care programs seem persuasive. Why are they not universally followed? How do these recommendations apply in the United States?

14. (This and the following are actual construction projects.) Make the ORT mixture according to Fig. 11.2 and taste it. Do you think the process shown in Fig. 11.2 to make the mixture would be effective? What further suggestions do you have to promote the use of ORT?

15. Make a model of a VIP latrine. Estimate the costs of a real one in the United States and how long it would take to build.

16. Build a water filter from materials locally available in your area, such as sand, and test it.

17. Design, build, and test a solar still.

# *12*

· · · · · · · · · ·

# SOLAR ENERGY

· · · · · · · · · · · · · · · · ·

An attached solar greenhouse can provide
solar heat for the house, a place to grow
vegetables and a pleasant spot to relax.

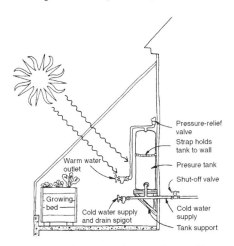

A tank placed in a solar greenhouse will
provide warm water.

## OVERVIEW

"100% Solar Home Goes Online", "Power from the Sun Will End Our Energy Woes" — headlines like these in popular magazines promise "free" energy. However, in typical homes today the promise is rarely kept. In this chapter we will try to separate the promises from the practical applications of solar energy.

Space heating, water heating, and direct conversion to electricity are all effective routes to study solar energy and important uses in practice. Using the sun to heat homes and workplaces is the most straightforward example for exploring solar technology. On cold winter mornings, one can usually find a comfortably warm, sunny spot in front of a south-facing window. Most summer mornings, one finds that same spot uncomfortably hot. By integrating some newer technologies into conventional building practices, that warm spot in winter has grown to include whole buildings. Through the same effort, the hot spot of summer has shrunk to bearable size.

## SOLAR SPACE HEATING

We examine the problem of keeping a home warm in the winter. Two processes determine indoor temperature: (i) the heat gained from the sun or the burning of fuel and (ii) the heat lost to the outside. Our goal is to maximize the heat gained and to minimize the heat lost. We maximize heat gained by efficiently capturing sunlight or burning fuel. We minimize the heat lost by reducing the ways and speed at which it travels to the outside.

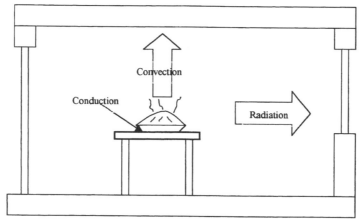

**Fig. 12.1. Heat transfer from a pie.**

To reduce our heat losses we need to understand how heat travels. Suppose we bake a pie (Fig. 12.1). When it is done we take it out of the oven and put it on a table to cool. After an hour we know that the pie will be cool enough to serve. What happened to the heat? It is gone, but where, and how?

*Fact*: It is the nature of things that heat flows from hot to cold.

It is a difference in temperature that drives heat from one place to another. Nature strives to keep everything at the same temperature. Therefore, the heat from the pie warmed the table, the air surrounding it, and the walls of the kitchen. If there are windows in the room some of the heat has gone through them. The amount of heat in the pie was small and the heat went to so many places that we may not notice much change in temperature in many of the objects that received the heat, although the table under the pie would be warm to the touch. To measure the increase in temperature of the walls or the air in the room would call for an extremely sensitive thermometer. Even if the temperature difference is very small, any difference in temperature will cause heat to flow from warm to cool.

*Fact*: Heat travels three ways: by conduction, convection, and radiation.

Our pie warmed the table by conduction. Conduction requires some medium or material through which the heat can travel. There is no perceptible motion of the medium itself. Convection took some of the heat from our pie and warmed the air in the kitchen. For convection to take place we need a fluid (gas or liquid). Unlike conduction, the motion of the medium carries the heat. As the pie warmed the air around it, that air, being lighter, rose, and cooler air took its place. This is called natural convection. You can see it in steam rising from a cup of coffee or smoke rising from a fire. If we serve the pie too hot, our guests may have to use forced convection (blowing) to cool it.

Radiation took heat from the pie out the window. In contrast to conduction and convection, radiation needs no medium, only a difference in temperature. Heat traveling by radiation warms our hands by a fire or burns us at the beach. But what does this pie knowledge have to do with our house? In our house the process is reversed; we do not want to lose heat, we want to keep it. We consider each of the ways that heat is transferred and determine what we can do to minimize the loss.

## Heat Loss through Conduction

Conductive heat loss occurs through the exterior walls, floors, and the roof — anywhere the exterior temperature is lower than the interior. Also, the rate of heat loss depends on the difference in temperature. The greater the difference, the faster the heat travels. The rate also depends on the material. Copper conducts heat rapidly, making it a good material for pots and pans but a poor one for walls. Dead (unmoving) air is a poor conductor. If we incorporate some dead airspace in our walls we reduce our heat loss. Insulation (fiberglass batting, cellulose, etc.) works using this principle.

To evaluate building materials, a measure of how quickly the materials transfer heat was developed. This measure is called the heat transfer coefficient ($k$). Table A.1 gives the heat transfer coefficients of some common materials. The coefficient $k$ is also called the thermal conductivity of the material.

In addition to the material and the temperature difference, the area and the thickness of the wall also affect the heat loss through a wall. A large wall area will conduct more heat than a small area. It takes heat longer to travel through a thick wall. When we calculate the heat lost through a wall during a day we must account for the thickness of the wall and its area.

We discuss a compact (mathematical) expression for the relation between these factors. First, some abbreviations:

$Q$, heat loss per second

$A$, area

$t$, thickness

$T_{in}$, indoor temperature

$T_{out}$ outside temperature

It was previously stated that as the area and the temperature difference increase the heat loss also increases. This can be written as

$$Q \propto A(T_{in} - T_{out})$$

where $\propto$ means "proportional to." As we increase the thickness, the heat loss decreases. So,

$$Q \propto \frac{A}{t}(T_{in} - T_{out})$$

What we want is for $Q$ to be equal to something, not just proportional. We also have not put in $k$, the heat transfer coefficient of the material.

Fortunately, the heat transfer coefficient takes care of both these problems. The equation used to evaluate conductive heat transfer is

$$Q \propto \frac{A \cdot k}{t}(T_{in} - T_{out}) \qquad [12.1]$$

To check this equation we examine the units attached to our variables. Heat is a form of energy so it should be measured in joules (see Chapter 2). $Q$ is the rate of heat loss so its units are joules per second. Actually, joules per seconds is equal to another unit, Watts. We measure temperature in degrees Celsius ($^\circ$C), area in square meters (m$^2$), and thickness in meters. When we add the units our equation is as follows:

$$Q(W) = (A(\text{m}^2)k/t(\text{m}))(T_{in} - T_{out})(^\circ\text{C})$$

To make the units on the right match Watts, the heat loss coefficient should be measured in Watts per meter per $^\circ$C.

When you consult a handbook to find values for heat transfer coefficients you will see they are tabulated in W/(m$^\circ$C) or equivalently Btu/hr/(ft$^\circ$F) units. The units of $k$ show why it is also called thermal conductivity. The quantity $k$ in English units is equal to the amount of heat conducted in 1 hr through a piece of material 1 ft thick and 1 ft$^2$ in area when the temperature difference between the two sides of the material is 1$^\circ$F.

For example, consider house insulation. A cool but reasonably comfortable interior temperature is 20$^\circ$C. On a chilly day the exterior temperature might be 2$^\circ$C. How much heat is lost in a 1 × 1-m block of fiberglass insulation 0.1 m thick? The value of $k$ is 0.033. Equation [12.1] shows that we would lose 5.94 W through the insulation. A block of steel of the same size, with a thermal conductivity of 54, would lose 9720 W. These examples show how much the rate of heat transfer is controlled by the material.

The reader may wonder why we have gone through this tedious derivation. The obvious reason is to find a way to estimate conductive heat loss. The important reason is to show that what we are doing is reasonable and consistent. If you take the time to analyze a problem, you can develop a way to solve it.

Recall that our original mission was to minimize the heat loss. If we look at the heat loss equation we can see that decreasing the area, decreasing the heat transfer coefficient, and/or increasing the wall thickness will decrease the heat loss. Therefore, the solution is obvious: Make the wall very thick and the house very small and from a material that has a small heat transfer coefficient. Unfortunately, here is where we run into reality. We must also weigh material cost and availability, construction cost, and reasonable building size into our solution. We will estimate our losses when we work out a practical example later in the chapter. Now, we explore another mechanism for heat loss.

## *Heat Loss through Convection*

Convection is the transfer of heat from solid surfaces by moving liquids and gases; our pie cooled faster because the air surrounding it moved up after being heated and the replacement air was cooler and therefore absorbed more heat from the pie. There are two types of convective transfer: free or natural convection and forced convection. Free convection is due to the fluid's change in density as its temperature changes. Air warmed over a hot surface, such as a radiator, becomes less dense and rises. Forced convection takes place when a fluid is forced, by a fan or pump, past a surface at a different temperature than that of the fluid. Forced convection is the mechanism of heat transfer in an automobile heater. A fan blows air over tubes heated by water from the car's engine. Forced convection can move more heat than free convection for a given temperature difference, which is why small electric space heaters have a fan.

The equation describing convective heat transfer is similar to that describing conductive heat transfer, but we use another coefficient instead of $k$ and there is no thickness to consider.

$$Q_c = h_c \cdot A(T_s - T_f) \qquad [12.2]$$

where

$Q_c$ is the rate of heat transfer

$h_c$ is the average convection heat transfer coefficient

$A$ is the surface area in contact with the fluid

$T_s$ is the surface temperature

$T_f$ is the fluid temperature.

For air in free convection the convective heat transfer coefficient is between 5 and 30 $W/(m^2 \cdot {}^\circ C)$ and about 3 $BTU/(s \cdot ft^2 \cdot {}^\circ F)$. Table A.2 in the Appendix to this chapter lists some convective heat transfer coefficients.

Why do we care? One example is heat loss from a swimming pool. A comfortable temperature for the water might be 25°C. The pool might be exposed to moving air at 10°C, corresponding to a cool breeze blowing over it. We will take the heat transfer coefficient to be 25 W/m$^2$/°C. Eq. [12.2] gives us the heat lost per second from each square meter of water surface:

$$q/A = 25(25-10) = 375 \text{ W/m}^2$$

The water surface area of a small swimming pool might be 10 × 5 m. In this case, the pool loses 18.75 kW to the wind — equivalent to about 187 light bulbs — which is why swimming pool heaters use much energy. The situation is actually worse because only about half the energy from a pool is lost by convection. Convective heat transfer is also important when estimating heat loss from a window or when calculating the efficiency of a solar collector.

# TOTAL HEAT LOSS FROM A BUILDING

We discuss a major calculation — the amount of heat lost in a building. We do this for several reasons: to estimate how much heat we must supply to keep the house comfortable and to understand where the heat is going so we can judge where to fix the house. Heat flows out of the walls, the roof, the windows, the door, the foundation, and through infiltration. In most cases only a small amount of heat flows out directly into the ground because dry earth is a good insulator, so we can disregard this heat loss. We must first examine the walls more carefully.

## *Heat Flow through a Typical Wall*

Most walls, instead of being made of a single material, usually are composed of several different materials forming a sandwich. A typical insulated wall for a house is illustrated in Fig. 12.2. The exterior is sheathed with wood 18 mm (0.75 in.) thick. The interior wall is plaster and the middle is filled with fiberglass. Each of the materials that make up the wall have different conductivities; therefore, we must combine the conductivities to find the total heat loss.

The effectiveness of a piece of insulation, whether a single material or a sandwich, is specified by an $R$ value, which includes both the conductivity and the thickness. To understand the $R$ value, we need to digress. The $R$ value is a material's resistance to conductive heat transfer and is defined as

$$R = \frac{\text{thickness}}{\text{thermal conductivity}} = \frac{t}{k} \qquad [12.3]$$

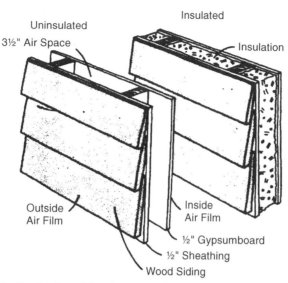

**Fig. 12.2. Typical residential wall construction. (*Energy Primer Solar, Water, Wind, and Biofuels*, Richard Merrill, Thomas Gage. Copyright Dell Publishing.)**

By using thermal resistance $R$ multimaterial calculations are easier to make because $R$ values add for the different material making up a wall. It makes sense for $R$ values to add because the $R$ for each material is directly proportional to the thickness of the material. To calculate the total thickness of a wall we would add the thicknesses of each layer of the wall. Heat flow depends not only on thickness but also on thermal conductivity; a thin material with a low thermal conductivity conducts heat poorly, acting like a thicker material with a higher heat conductivity. Therefore, to find the total effective thickness, analogous to resistance, we add the effective thicknesses of the layers, with these effective thicknesses being the actual thickness weighted by the conductivity.

To express heat transfer in terms of the resistance $R$ we combine our original equation for heat transfer (Eq. [12.1]) with the equation for $R$. Equation [12.1] is rewritten here for convenience.

$$Q = \frac{A \cdot k}{t}(T_{in} - T_{out})$$

By dividing Eq. [12.3] into Eq. [12.1], the equation for heat transfer becomes

$$Q = \frac{A}{R}(T_{in} - T_{out}) \qquad [12.4]$$

### Table 12.1a. Thermal Resistance in Metric Units

| Material | Thickness (m) | $k$ ($W/m/^\circ C$) | $R$ ($m^2/^\circ C/W$) |
|----------|---------------|----------------------|------------------------|
| Wood | 0.018 | 0.147 | 0.164 |
| Fiberglass | 0.090 | 0.038 | 2.37 |
| Plaster | 0.012 | 0.47 | 0.026 |

### Table 12.1b. Thermal Resistance in English Units

| Material | Thickness (in.) | $k$ ($Btu\ in./hr/ft^2/^\circ F$) | $R$ ($hr/ft^2/^\circ F/Btu$) |
|----------|-----------------|-----------------------------------|------------------------------|
| Wood | 0.75 | 0.8 | 0.937 |
| Fiberglass | 3.5 | 0.27 | 12.69 |
| Plaster | 0.5 | 8.0 | 0.062 |

The total resistance $R$ in Eq. [12.4] is the sum of the $R$s for the material forming the walls. Now we can go back to our house example. First, we will calculate the $R$s for the various materials using metric units. Values for $k$ and $R$ can be found in handbooks (see References). These have been copied in Table 12.1a.

Using the values in Table 12.1a, the value of $R$ for the wall is

$$R \text{ (wood)} + R \text{ (fiberglass)} + R \text{ (plaster)} = 0.164 + 2.37 + 0.026 = 2.56$$

Note that in this wall the fiberglass blocks the heat most effectively. If the temperature difference is $18^\circ C$, then the heat loss, from Eq. [12.3], is $7.03 \text{ W/m}^2$.

Using the English system of units, the thickness, the value of $k$, and the value of $R$ are shown in Table 12.1b. Therefore,

$$R = 0.937 + 12.69 + 0.062 = 13.689$$

With a temperature difference of $32^\circ F$ we have a heat loss of $2.34 \text{ Btu/hr/ft}^2$.

## CALCULATION OF HEAT LOSS FOR A SIMPLE HOUSE

We consider a very simple house (Fig. 12.3). The back wall has no windows and the other side wall is identical to the one shown. We estimate the heat flow out of the house in Btu/hr when the outside temperature is 32°F (freezing) and the inside temperature is 65°F (about as cool as we would like it).

Heat, as described previously, flows out the roof, the windows, the door, and the foundation and through infiltration (heated air escaping through cracks). Infiltration is a special problem that we will examine after we calculate the conductive heat losses. Heat loss through the foundation is also somewhat different because most of the heat flows through the top part of the foundation so the length of the foundation, not its area, is what determines the heat loss. The length of the foundation is the perimeter of the house. The $R$ value for the foundation is given in terms of feet, not

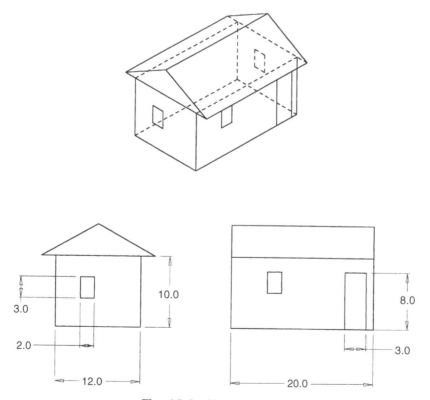

**Fig. 12.3. Simple house.**

**Table 12.2. Typical *R* Values**

| Material | R value |
|---|---|
| Walls | 13.7 |
| Roof (with 6 in. fiberglass batting) | 25 |
| Window (single glazing) | 0.88 |
| Window (double glazing) | 1.54 |
| Door | 2.3 |
| Foundation (with foam insulation) | 20 |

square feet. We might start by looking up *R* values for the materials used in the house construction. To simplify matters, this research has been done and the results are shown in Table 12.2. From these *R* values we will estimate the heat losses.

*Walls*: The *R* number given in Table 12.2 assumes the walls are made as shown in Fig. 12.2. Equation [12.4] can be used but we must find the area of the walls first.

Front: $12 \times 20 - (8 \times 3) - (2 \times 3) = 210 \text{ ft}^2$

Sides: $2 \times (12 \times 12 - (2 \times 3)) = 276 \text{ ft}^2$

Back: $12 \times 20 = 240 \text{ ft}^2$

   Total: $726 \text{ ft}^2$

Using Eq. [12.4] we have:

$$Q_{\text{Walls}} = \frac{A}{R}(T_{\text{in}} - T_{\text{out}}) = \frac{276 \cdot 33}{13.7} = 665 \text{ Btu/hr}$$

*Roof*: The calculation is the same. The area is $240 \text{ ft}^2$

$$Q_{\text{Roof}} = 316 \text{ Btu/hr}$$

*Windows*: We assume single glazing has been chosen for all three windows. It should be noted that convection plays an important role in heat transfer through windows in the winter because air flows up along the inside of the window as it is warmed. This convection was considered when the *R* values shown in Table 12.2 were calculated.

$$Q_{\text{Windows}} = 675 \text{ Btu/hr}$$

*Door*:

$$Q_{\text{Door}} = 343 \text{ Btu/hr}$$

Of course, this value of $Q$ does not account for the large heat loss when the door is open.

*Foundation*: Heat loss through the foundation is calculated the same way except we use the length of the foundation, the perimeter of the house. The perimeter of the house is 64 ft:

$$Q_{\text{Foundation}} = \frac{64 \cdot 33}{20} = 105 \text{ Btu/hr}$$

*Infiltration*: Cold outside air blows into the house through cracks and this air must be heated to make the house comfortable. If air did not blow in, the house would be uncomfortable for another reason — buildup of odors and $CO_2$. Less than one air change every 2 hours, or 0.5 air changes an hour, will make the house uncomfortable. We assume that in every hour half the air in the house must be heated from 32° to 65°F. It takes 0.018 Btus to raise 1 cubic foot of air 1°F. (This number is called the specific heat.).

$$Q_{\text{Infiltration}} = \text{volume} \cdot \text{specific heat} \cdot \text{temperature rise} \cdot \text{number of air changes}$$

$$Q_{\text{Infiltration}} = (12 \cdot 12 \cdot 20) \cdot (0.018) \cdot 33 \cdot 0.5 = 855.36 \text{ Btu/hr}$$

The total heat loss is found by adding all these losses:

$$Q_{\text{Total}} = Q_{\text{Walls}} + Q_{\text{Roof}} + Q_{\text{Windows}} + Q_{\text{Door}} + Q_{\text{Foundation}} + Q_{\text{Infiltration}}$$
$$= 4044 \text{ Btu/hr}$$

Why do we care about this calculation of heat loss? It shows us how we can most effectively reduce that loss. In our example, it makes much more sense to reduce losses through the windows than through the roof. Also, the calculation shows us how much heat we must bring into our house every hour. Actually, if we are using solar heat, we can bring heat in only when the sun is shining so we must be able to store heat. To determine how much heat we need to bring in and store we need to calculate our daily heat losses. To find our daily heat loss, we multiply our average hourly loss by 24. To find the average hourly loss we use an average daily outside temperature. In our case, if we assume the average outside temperature cycles between 28° and 36°F and we want a constant 65°F inside, then we can use, as we did previously, 32°F as our outside temperature and 33°F as the temperature difference. In this case the total daily heat loss is

$$Q_{\text{Daily}} = Q_{\text{Total}} \cdot 24 = 70{,}784 \text{ Btu/day} \qquad [12.5]$$

Actually, we do not need to supply all this heat because the occupants and electrical appliances will supply some. A person produces about 250 Btu/hr and a large dog 200 Btu/hr. Electrical appliances, such as lights or toasters, also supply heat; 1 W/hr = 3.4 Btu. Therefore, a 100-W bulb on for 10 hours supplies 3400 Btu. If our house held 1 person and a dog and used 200 W for 10 hours, the energy supplied would be

$$Q_{\text{Supplied}} = (250 \cdot 24) + (200 \cdot 24) + (200 \cdot 10 \cdot 3.4) = 17{,}600 \text{ Btu/hr}$$

So the additional heat needed is

$$Q_{\text{Needed}} = Q_{\text{Daily}} - Q_{\text{Supplied}} = 53{,}184 \text{ Btu/day} \qquad [12.6]$$

In the next section we consider how to get this much heat from the sun. We start by examining a third way heat travels — radiation.

## HEAT RADIATION

Most of us have seen a greenhouse or cold frame and know, that in bright sun, it is warmer inside than outside. Why is this? Why does putting a glass cover over something allow it to be warmer? To answer this we need to understand how heat radiates. Suppose we heat a steel bar in a furnace. As it gets hotter it begins to glow; first, dark red, then orange, then yellow, and finally almost white — too bright to look at. The glow we see is energy being radiated. The sun is a perfect example of something glowing "white hot" and radiating energy. In an attempt to analyze the behavior of radiated heat (and light), scientists developed the theory of electromagnetic radiation. A central idea of this theory is that radiation travels in waves. If it were possible to isolate and make visible a single wave from the many waves around us, the single wave would look like that shown in Fig. 12.4. As you can see there are two quantities that describe the wave: amplitude and wavelength. Amplitude gives a number to the intensity of the heat or light. Wavelength gives a number to the "color" of the wave. The color violet (the shortest wavelength people can see) has a wavelength of about 0.4 μm (1 μm = $10^{-6}$ m). Red (the longest visible wavelength) has a wavelength of about 0.6 μm. Thermal radiation ranges from about 1000 to 0.25 μm.

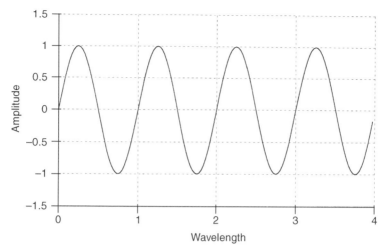

**Fig. 12.4. Typical wave.**

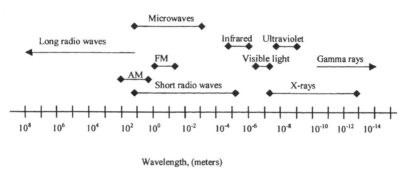

Wavelength, (meters)

**Fig. 12.5. Electromagnetic spectrum.**

Electromagnetic radiation exists in a broad spectrum of wavelengths. Figure 12.5 shows the names given to various parts of the spectrum. Energy from the sun is concentrated in the visible wavelengths. This means that most of the sun's heat and light comes in these wavelengths. Figure 12.6 shows the spectrum of sunlight — how much of the sun's energy is at each frequency.

Fortunately for our understanding, heat radiation behaves very much like light; it can be reflected, transmitted, absorbed, and emitted. Our experience with light gives us an intuitive understanding of heat radiation. Recall the steel bar in the furnace: Before it glowed it emitted radiation at wavelengths longer than we can see. The wavelength of the radiation that a body emits depends on its temperature; the higher the temperature, the shorter the wavelength. The wavelength also depends on the emitter

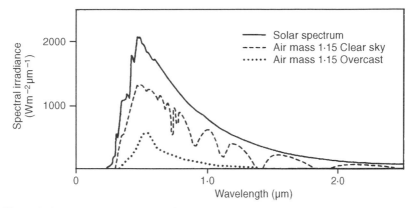

**Fig. 12.6. Solar spectrum. (***An Atlas of Renewable Energy Resources*, **Julian Mustoe. Copyright John Wiley & Sons Limited. Reproduced with permission.)**

material and its surface (polished, dull, rough, etc.). Materials respond differently to different wavelengths.

The interior of a greenhouse becomes warm when sunlight hits it because of the differences in the response of the glass to different wavelengths. The glass of the greenhouse is transparent to the wavelengths in sunlight. The glass, however, absorbs — retains rather than transmits — radiation of longer (and shorter) wavelengths. The plants and soil inside the greenhouse, because they are cooler than the sun, radiate heat at longer wavelengths. Thus, the heat from the sun passes through the glass, is absorbed by the plants and soil, and is reradiated. This reradiated energy, because of its longer wavelength, does not pass through the glass but is absorbed and reradiated again. Some of the reradiated energy goes to the inside of the house and some to the outside. For a typical greenhouse about 83% of the sun's heat is transmitted through the glass but only about 0.003% of the radiation from the interior passes back through the glass.

What happens when thermal radiation hits an object? There are three possibilities: It may be transmitted, reflected, or absorbed. It was shown with glass that what happens is dependent on the wavelength of the radiation: Visible light is transmitted and thermal radiation is absorbed. What happens also depends on the surface — a rough surface will absorb more energy than a shiny one — and on the internal absorption of the material. High internal absorption means the material is opaque to the incident radiation. Low absorption means that the material is transparent. Again, glass is a good example. Impurities can raise the internal absorption and make the glass opaque. Some glass is made with a rough surface so it transmits a small fraction of the incident light.

## SOLAR RADIATION

To estimate how much heat we can actually get from the sun we need to know how much energy the sun radiates or, more precisely, how much energy is received from the sun on the earth's surface. This has been measured. The radiant power reaching earth's outer atmosphere varies between 1.32 kW/m$^2$ in early July to 1.42 kW/m$^2$ in early January; the sun is actually closest to the earth in January. The solar constant is defined as the average power reaching a surface, outside the earth's atmosphere, 1 m$^2$ in area facing the sun directly. (Facing the sun directly means the surface is perpendicular to the solar radiation.) The solar constant is 1.36 kW/m$^2$. The equivalent number in "English" units is about 430 Btu/(hr·ft$^2$).

All this energy will not be available for heating. First, some of the energy will be lost — reflected or absorbed — as it goes through the earth's atmosphere. Second to collect the maximum amount of radiation the collecting surface must face the sun directly. It is convenient in making solar designs to examine four separate factors which determine the amount of solar energy received: geographic location, collector orientation, time of day and year, and atmospheric conditions.

### *Geographic Location*

The geographic location, essentially the distance from the equator, determines the amount of atmosphere the radiation must pass through before hitting the collector. The greater the latitude the more atmosphere the

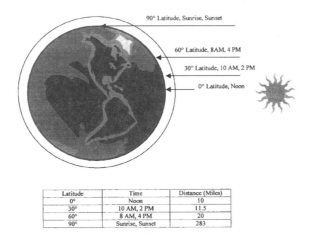

| Latitude | Time | Distance (Miles) |
| --- | --- | --- |
| 0° | Noon | 10 |
| 30° | 10 AM, 2 PM | 11.5 |
| 60° | 8 AM, 4 PM | 20 |
| 90° | Sunrise, Sunset | 283 |

**Fig. 12.7. Latitude and path through atmosphere.**

radiation must go through (Fig. 12.7). More atmosphere traversed results in more energy lost. The distance from the sun is also slightly longer further from the equator, but the sun is so far away that this increase in distance is insignificant. Location also determines the weather patterns. Obviously, a cloudy climate will be less productive than a sunny one.

## *Collector Orientation*

The more perpendicular the face of the collector is to the sun's rays the more radiation the collector intercepts (Fig. 12.8). Another way of visualizing the effect of collector orientation is to realize that the larger the shadow

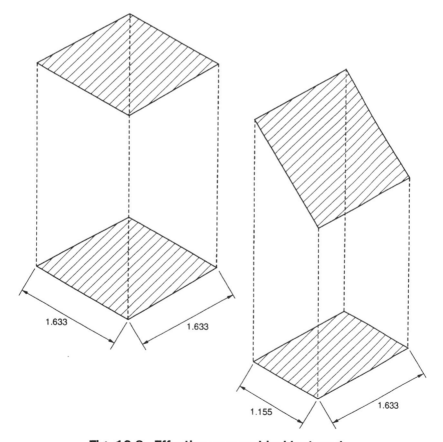

**Fig. 12.8. Effective area and incident angle.**

cast for a given area of collector the greater the radiation incident on the collector.

## *Time of Day and Year*

During the day the insolation (incident solar radiation) at a particular location reaches its peak at solar noon with minimums at sunrise and sunset. This is because at noon the location is facing the sun directly, whereas at sunrise and sunset the location is facing 90° from the direction of the sun. (A drawing such as that in Fig. 12.7 would show this change in collected radiation as the earth rotates and the reader might want to make the sketch.) Not only is the direction of the radiation best at noon but also the radiation has to go through less atmosphere at this time. Again, the idea is illustrated in Fig. 12.7.

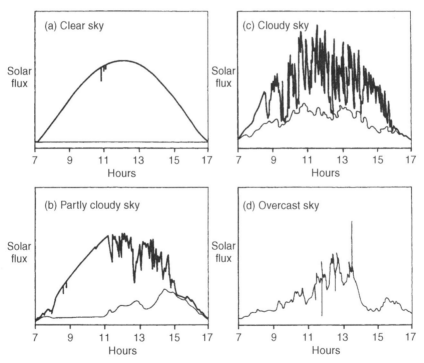

**Fig. 12.9. Insolation with respect to time of day and time of year. (*An Atlas of Renewable Energy Resources*, Julian Mustoe. Copyright John Wiley & Sons Limited. Reproduced with permission.)**

There is more solar energy available to collectors during the summer because the days are longer. The collector also faces the sun more directly. Also, seasonal weather patterns affect the radiation reaching the collector and summer weather tends to have fewer clouds. Figure 12.9 shows the change in insolation with respect to time of day and time of year. The dips in the graphs are clouds passing.

## Atmospheric Conditions

Cloud cover is the most important condition affecting the amount of radiation reaching the earth's surface, house, or solar collector. Clouds reduce incoming radiation by as much as 90% because they reflect that amount of radiation back into space. Ozone absorbs much energy in the infrared region. Carbon dioxide, oxygen, other gases, and dust all absorb radiation to some degree. The atmosphere scatters radiation as well as absorbing and reflecting it. Scattering is wavelength dependent and affects short wavelengths most. The reason the sky is blue is that short wavelengths are seen as blue, and these wavelengths are the ones that are scattered.

Global warming, which is discussed often in newspapers and magazines, is a result of the reflection of thermal radiation by cloud cover and other gases in the atmosphere. In this case the concern is with thermal radiation produced on the earth and reflected back to the earth's surface, warming that surface. A major problem is suspected to be carbon dioxide, a gas produced when burning wood, coal, oil, and other organic fuels. Increased amounts of carbon dioxide in the atmosphere will increase the amount of heat reflected back, warming the earth. This warming will produce significant changes in climate. One forecasted change is the melting of the ice in the Arctic and Antarctic, which will raise the level of the oceans and flood coastal areas.

The net effect of the atmosphere is to reduce the energy actually available on a flat surface on the ground, facing the sun directly, from 1.36 to about 1.0 $kW/m^2$ or from 430 to about 300 $Btu/(hr \cdot ft^2)$.

How can we parlay this information into something useful such as designing a house to be heated by solar energy? Let's think of the things necessary to heat our home successfully from the sun. Before we can use the heat we have to capture it. We'll need a surface that is a good absorber of solar wavelengths. A good absorber has low reflectivity and emissivity (emissivity is a measure of how readily the surface radiates energy). What do you do when the sun goes down or on a cloudy day? Obviously, we need some way of storing the energy.

# Passive Solar Heating

In designing a solar-heated house we must consider how the solar energy is collected, absorbed, stored, and distributed. We must also consider how we will regulate the temperature of the house, both in summer and in winter. In a typical solar system, radiation enters the house through large south-facing windows to be absorbed and stored by masonry walls and floors, water-filled containers, or rock beds under the house. Natural convection then distributes the heat to the rest of the living areas. A passive design is one in which the heat flows naturally through the space, in contrast to an active system, in which a fluid is heated and is then pumped through the house.

In a passive solar system both the design and the materials are chosen to make best use of the solar radiation. Because a passive solar system is integrated into the building rather than being a separate system (like conventional heating and cooling) its incorporation requires an integrated approach to the design. We have been thinking about heating the house in winter but we should also think about cooling. In summer, passive solar design minimizes the amount of sunlight entering the house and provides ventilation. In general, siting, landscaping, floor plan, window placement, and building materials all affect the performance of a passive solar-heated house.

## *The Five Components of a Passive Solar System*

A passive solar system will combine five interrelated components: collector, absorber, storage, distribution, and regulation. We will discuss these separately.

The collector is the portion of the building that actually lets in the sunlight. A collector can be south-facing windows, skylights, or panels especially made to collect and absorb. To be beneficial, collectors should be oriented to the south within 30°. They should be tilted relative to the earth's surface to face the sun directly. The angle of tilt should be approximately the latitude of their location, as Figs. 12.7 and 12.8 show. (Latitude is 0° at the equator and 90° at the poles.)

The absorber is the recipient of the solar radiation. It is warmed by the radiated heat from the sun and makes that heat available for direct heating or storage. Absorbers are usually painted black because black surfaces have the lowest emissivity of any color and so reradiate the least heat. Special paints are available with very low emissivity but these cost more than ordinary paints.

Storage is usually done in dense materials such as brick, rock, water, adobe, and concrete. Storage may be incorporated in a wall, floor, or room divider. A masonry or slate floor in the room with south-facing windows would be an effective storage medium, as would the back wall of the room. In these cases the surface of the floor or wall is the absorber. The storage needs to be of the proper size to hold sufficient heat for the particular use; a slate floor might be chosen that is thick enough to keep the room warm until midnight. A rule of thumb for residential spaces is 1 ft$^3$ of masonry for each square foot of window area. Because it is difficult to move heat a long distance by natural convection, storage is usually placed adjacent to the rooms it is intended to heat.

The distribution system moves the heat from the storage area to the rest of the building. Distribution is, in most houses, by natural convection. The designer can install passages for naturally flowing air currents so the warm air goes into the living areas. Occasionally, small fans or pumps are used to augment natural convection.

Regulation in a passive solar system is usually done by adjustable vents that can control the movement of heated air through the building. Another type of control device is insulation to cover windows at night to prevent stored heat from escaping. Regulation is more difficult in a passive system than in a conventional heating system with automatically controlled blowers to force the heat around; of course, the automatic system is more expensive.

## Direct Gain

Three basic techniques are used in passive solar systems: direct gain, indirect gain, and isolated gain. We consider direct gain first. In a direct gain system, sunlight goes directly into the space to be warmed, for example, solar radiation entering a room through south-facing glazing. The radiation is absorbed and stored in masonry walls and floors and any other masses that are directly warmed by the sun. Direct gain warms the living space immediately when the sun shines and provides good natural lighting and an outdoor view. The major drawback of direct gain is that it sometimes works too well and overheats the living space. Because storage is in the same place as where the heat is used, one has less control over the temperature than if the heat were stored separately from where it is used. Figure 12.10 illustrates direct gain.

We now estimate the size of the window for the house shown in Fig. 12.3. As mentioned previously, about 300 Btu come from the sun, through the atmosphere, each hour per square foot. We need 53,184 Btus in a day. The room collects at least some sun for about 8 hours a day, from 8 AM to

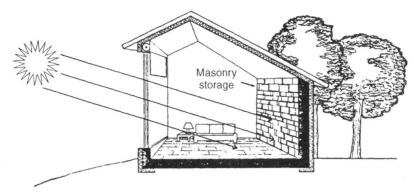

**Fig. 12.10. Direct gain.**

4 PM. Not all of the sun's energy gets through the glass; some which gets into the room is reflected back out the window, and some is absorbed by the carpet and draperies which do not radiate heat efficiently. Actually, only about 20% of the peak energy—the energy available at solar noon—outside the window actually heats the house [1,3]. The area of the window is calculated as follows:

$$Q_{In} = Q_{Daily}$$
$$Q_{In} = 300 \cdot \text{area} \cdot 8 \text{ hr} \cdot 0.2$$
$$Q_{Daily} = 53{,}184 \text{ Btu} \quad (\text{Eq. } [12.6])$$
$$\text{Area} = 111 \text{ ft}^2$$

[12.7]

Actually, this is the area perpendicular to the sun's rays, so a somewhat larger window would be useful, perhaps $10 \times 14$ ft.

Such a window would increase heat loss so double glazing would certainly be necessary. In a serious design we would recalculate our heat loss through these larger double-glazed windows to determine if we would have to make them even larger to make up for the additional loss. We must also provide for insulation at night over the window, which probably means the owner must remember to put up that insulation each night. The heat input would be less, perhaps half, on cloudy days so a backup heater, perhaps a woodstove, is also needed.

Sufficient heat storage is also necessary. Using our rule of thumb of 1 cubic foot of masonry per square foot of window and using masonry—floor and back wall—6 in. thick we calculate that a total area of 280 ft$^2$ is needed. This area might come from a floor and back wall $16 \times 18$ ft—a reasonable size.

## Indirect Solar Gain

Indirect gain collects radiation through the same south-facing glazing as direct gain but the absorber and the storage are separate from the living area. A commonly employed indirect gain system is an 8- to 16-in.-thick masonry wall built as the south wall of the house. The exterior surface of the wall is painted black to absorb the radiation and the entire wall is glazed—that is, a large glass window is between the wall and the outdoors. This wall is known as a Trombe wall (Fig. 12.11).

A Trombe wall provides heat in two ways. The first way is through convection. Two sets of vents are installed in the wall; one set at the floor level and another at the ceiling. When the sun shines on the wall the air in the cavity between the wall and the glazing is heated and rises, flowing through the ceiling vents. Cool air from the adjoining room is drawn in through the floor vents. The path taken by the air is called a natural convection loop.

The Trombe wall also provides heat by absorbing solar radiation and storing it. The wall is warmed by the sun and reradiates the heat to the rest of the house as long as the temperature of the wall is greater than that of the living space. The stored heat allows the wall to continue to provide heat after the sun has set. This time lag gives indirect gain systems the advantage of supplying heat when it is needed without making the living space uncomfortably warm.

## Isolated Gain

In an isolated gain system solar radiation is captured in a separate, glazed, otherwise unheated space and then transferred to the living space. Atriums

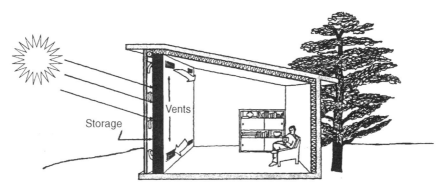

**Fig. 12.11. Indirect gain, Trombe wall.**

and attached greenhouses are examples of isolated gain. Solar greenhouses are the most practical method for retrofitting existing housing with passive solar heating and are the usual way of employing isolated gain. A greenhouse is advantageous for three reasons. First, the solar heat can be stored without having unsightly or bulky masonry or water tanks in the living areas and the heat can be distributed effectively to where it is needed. This distribution occurs by vents and ducts. Second, the greenhouse provides additional living space. Third, the greenhouse can serve as a buffer between the living space and the outdoor, thus reducing heat losses. The book by Shapiro [3] describes how to design attached greenhouses. An example of isolated gain is shown in Fig. 12.12.

## Cooling

Although our main objective in passive solar design is providing heat to a house, to be livable, it will also require some type of cooling. The effectiveness of passive cooling depends on how well heat gain is controlled and on adequate ventilation. Ventilation removes heat generated inside the house, for example, by the people living there. Moving air also feels more comfortable than still air. Heat can be kept out by shading the collection area with overhangs or vegetation. Moveable insulation can be used to reduce heating through windows. House design can promote natural ventilation. Such design requires a clear path for summer breezes and use of the chimney effect to pull cooler air into the house. The chimney effect is produced by warm air rising and flowing out of the house, as smoke rises and flows out of a chimney. Traditional housing in the Middle East and in West Africa has made remarkably effective use of natural ventilation for centuries.

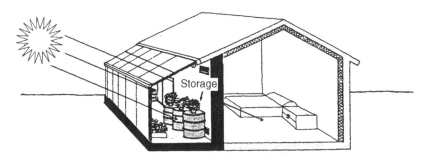

**Fig. 12.12. Isolated gain.**

## Siting a House

Careful placement of a house will make passive solar heating more effective. The ideal site for a passive solar home is on a south-facing slope with evergreen trees to the north of the house and deciduous trees in front. By building the house into the slope protection is provided from winter winds on the back, whereas the front of the house has access to the sun. If the site is flat the same effect can be accomplished by building up the earth on the north and sometimes east and west sides. This partial burying of the house is called earth berming and protects the house from air infiltration and smooths out the temperature extremes. A further extension of berming is to bury the house into the hill.

The amount of water in the soil also influences the effectiveness of solar heating. A low water table and good drainage will reduce the loss of heat to the soil. Wet soil will cause damp basements and floor slabs and will conduct heat away from the house much faster than dry soil. Soil is a heat sponge when it is wet; it is a fairly effective insulator when dry. Not much can be done to improve a damp site; these should be avoided.

Trees do more than simply add beauty. The size, location, and number of existing trees is important when one is selecting a site — waiting for new trees to grow can take a long time. A healthy stand of evergreens to the north, northeast, and northwest will provide a much needed windbreak. Deciduous trees to the south can shade the collector area during warmer weather. It may be possible to plant deciduous trees in such a way as to channel summer winds into the house. In general, topography, soil, drainage, and vegetation must be considered to make the best design possible.

## Environmental Factors

In our design calculations so far we have simply assumed a particular outside temperature and availability of sunlight. These temperature and sunlight data would probably be the averages expected at the site. In a more thorough design we would consider how much and how long conditions varied from the average. The available sunlight, the number and pattern of sequence of cloudy days, the direction and speed of the prevailing wind, the average, maximum, and minimum daily temperatures, and the relative humidity are all factors that must be taken into account when proceeding with a passive solar design. (High humidity can decrease the comfort level on both hot and cold days.) Climatic data about a particular site can be found in a number of sources, some of which are given in the References.

The number of degree days is often mentioned when describing climatic conditions at a particular site. Degree days are accumulated daily throughout a heating season by subtracting the average outdoor temperature from a base of 65°F. For example, one day with an average outdoor temperature of 20°F produces 45 degree days, 10 such days produce 450 degree days. A house in Caribou, Maine, will not have the same requirements as a house in Oak Ridge, Tennessee. In Caribou, the heating demand is 9767 degree days. In Oak Ridge the heating demand is 3800 degree days. The outside temperature in Oak Ridge is closer to 65°F and for more days.

The primary function of the design for a house in Caribou, based on the number of degree days, will be to provide adequate heat for a long cold winter. In Oak Ridge the goal will be to provide a balance between cooling for the summer and heating for a relatively mild winter. A major factor in determining how passive solar design is used is the climate.

## Solar Access

Passive solar design requires adequate access to sunlight. Usually this is not difficult for single-family houses but multifamily units and closely spaced developments will need careful planning so each unit receives sun. North of the equator, the winter sun is never overhead. It is always in the southern sky. A solar house ideally should face south and, if it cannot face due south, should always face within 30° of south. During the peak radiation period from 9 AM to 3 PM the collection area should not be shaded by trees, hills, or other buildings. Sunlight collected before 9 AM and after 3 PM is also of value but will come from the east or west, so if one wants to collect the early morning or late afternoon sun one must have east- or west-facing windows. Early morning sunlight, for example, can provide quick heat for an east-facing breakfast area. The house designer needs to work out the solar window — the portion of the sky from which the house receives sunlight — and determine if a significant portion of that solar window is blocked and whether the house will receive sufficient sun. Astronomical data given in the References are the basis for determining the solar window.

## Energy Conservation

An important part of a passive solar design is reducing the heat loss from the house. Insulation and double glazing, as described previously, will certainly reduce the heat loss and a reduction in the heat loss will in turn reduce the size of the collection system. The design of the house also influences the heat loss. The major issues for a passive solar house are the shape and orientation of the building and the layout of the interior rooms. The shape of the building determines its solar gain and heat loss. The ideal shape would consist of a large area facing south, filled with windows, and a small north-facing area that is well insulated.

The orientation of the house should be chosen so that the long axis of the house runs east to west, allowing a south-facing solar collection area that is large enough to be shared with most of the living space. The orientation of this south-facing wall must be chosen to collect solar radiation effectively, i.e., within 30° of true south.

Rooms should be laid out so as to reduce heat loss and take advantage of solar gain and natural convection. A good design situates the low-use and unheated areas, such as the garage and utility room, on the north side of the building, thus making a buffer between the heated living space and the north wall. Bathrooms, with small windows, also can be placed on the north side. The kitchen might be placed on the east end of the house, where it will catch the morning sun.

To reduce heat loss it is necessary to insulate and seal the building envelope as discussed previously. Maintaining a low exterior surface: interior volume ratio and building a compact structure improves heat conservation. A major heat sink is the leakage of air through cracks around doors and windows and through joints in the walls and roof (infiltration). Weather stripping and caulking will minimize these leaks. A vapor barrier will also help reduce infiltration. It should be cautioned, however, that if a house is made too tight, condensation and stagnant air will result. A rate of 0.5 air changes per hour is recommended. Eliminating or minimizing the windows on the north side and using double and triple glazing elsewhere will decrease the losses through windows.

Walls, ceiling, and floor all lose heat by conduction. The proper installation of insulation in walls, floor, crawl space, roof, and foundation will reduce this loss to acceptable levels. Another area to be addressed in the design phase is the entrance. Every time an outside door is opened and closed cold air from the outside is exchanged for warm air from the inside. An "air-lock" or unheated vestibule that creates a two door entry will allow only the cool air in the entry to be exchanged.

# SOLAR HOT WATER SYSTEMS

The next few sections discuss the design and operation of a flat plate hot water heating system. The discussion focuses on flat plates, systems, and economics. A flat plate collector could be used to heat a house but it is usually cheaper to use direct, indirect, or isolated gain for space heating rather than build a collector.

The most visible evidence of solar energy conversion to a casual observer in the United States or abroad is the flat plate water heater. In the United States, 1 in 50 homes has some type of flat plate collector installed. In Greece and Israel solar hot water systems are common. This is not to say that all these are operational; a maintenance-free system has not been perfected.

The essential parts of a flat bed collector are an absorber to collect heat and fluid to transfer heat from the absorber to where it is useful. The sophistication of the design of flat plate collectors ranges from trickling water down a sheet of corrugated steel into a trough to a triple glazed all-copper absorber with optimized coating. The design chosen depends on the application, economics, climate, and required performance.

### *Trickle-Type Collector*

The trickle-type collector uses a sheet of corrugated metal that is painted black and tilted toward the sun. Once the sun warms the plate, water is

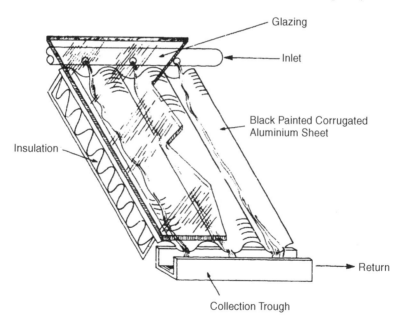

Fig. 12.13. Trickle-type hot water heater. (*More Other Homes and Garbage*, Jim Leckie, Gill Masters, Harry Whitehouse, Lily Young. Copyright Sierra Club Books.)

trickled from a tube at the top of the collector. The water is heated by the plate as it flows down into a trough at the bottom. The warm water in the trough is piped to storage. Figure 12.13 illustrates a typical trickle collector. Glazing (the glass or plastic that covers a collector) is often omitted from this type of collector because the heat loss by radiation from the heated water is not significant. The main advantage of a trickle collector is its ease of construction and economy. An unglazed trickle collector could be built for less than $20.

## Tube-in-Plate Collectors

As the performance requirements of the application increase, so does the complexity of the collector. More prevalent than trickle collectors are "tube-in-plate" absorbers. A generic tube-in-plate is shown in Fig. 12.14. In this design, the fluid-carrying tube is incorporated in the absorber plate. This offers several immediate advantages over the trickle collector. The water is not exposed to airborne contamination, there is no evaporation, and the fluid is warmed by a larger heated surface.

Several companies offer absorber plates that have tubes formed in them during the manufacturing process. User-built plates may have tubes bonded to plates by soldering or some mechanical fastener.

The efficiency of a collector is determined by the thermal conductivity of the tube and plate material, the losses to the surroundings, the ability of

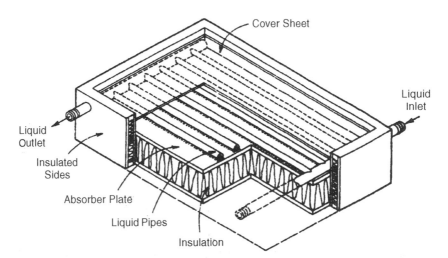

**Fig. 12.14. Tube-in-plate collector. (*Solar Energy Fundamentals and Design*, William Stine and Raymond Harrington.)**

the plate to absorb solar energy, and the path of the fluid through the plate. An absorber would be most efficient if its temperature were constant over the whole surface; efficiency is greater when the temperature of the material losing heat and the material gaining heat are nearly equal. Obviously, the temperature cannot be the same all over the plate if we are inserting cold water in one side and taking warm water out the other. Efficiency is improved, however, if the input water warms up quickly as it flows through the collector.

### Materials

The choice of tube and plate material is a trade-off between thermal characteristics, cost, and joining. An added constraint is that the tube material not corrode nor contaminate the water. This constraint limits the tube selection to copper or plastic, materials used in plumbing. Table 12.3 provides some properties for a variety of materials.

We now consider the implications of the several parameters in Table 12.3. Thermal conductivity is how fast a material transfers heat from a hot region to a cooler one. The higher the thermal conductivity, the faster heat is transferred to the incoming fluid, the faster that fluid warms, the more uniform the temperature, and therefore the more efficient the collector.

The second parameter in Table 12.3 is the cost of the material, which determines the initial cost of the project. If the initial cost is too high the project may not be possible for the homeowner. However, the homeowner does need to consider long-term benefits and problems of the project in addition to the initial cost. In many cases a solar hot water heater will pay for itself within 6 years. Copper tubing is expensive but is rugged and easy to solder.

The last two parameters in Table 12.3 measure corrosion resistance and thermal expansion. Corrosion resistance determines life of the panel

### Table 12.3. Material Properties

| Material | $k$ (W/(m°C)) | Cost | Corrosion resistance | Coefficient of thermal expansion | Availability |
|---|---|---|---|---|---|
| Silver | 407 | Highest | Good | $1.84 \times 10^{-5}$ | Poor |
| Copper | 386 | High | Good | $1.62 \times 10^{-5}$ | Excellent |
| Aluminum | 204 | Medium | Fair | $2.22 \times 10^{-5}$ | Good |
| Steel | 54 | Medium | Poor | $1.14 \times 10^{-5}$ | Good |
| Plastic | 0.195 | Low | Excellent | $6.75 \times 10^{-5}$ | Good |

and whether the tube will contaminate the fluid. The coefficient of thermal expansion is included in the table because absorbers require some type of bond between the tube and the plate. For good thermal performance this area should be maximized. If the tube has a coefficient of thermal expansion very different than that of the plate the two will expand at different rates and may break the bond.

The implication of the information in Table 12.3 is that copper plates and tubes are most often used. Although their cost is higher than that of the other materials shown (except for silver, which is prohibitively costly), the improved efficiency and the ease of construction usually outweigh the cost. Second most common in usage is plastic, used mostly in applications in which economy is most important.

### Construction of the Absorber

Two design decisions are made in constructing the absorber part of the collector. The first is the path the fluid takes in going from input to output. The second is what kind of a coating to put on the absorber surface.

The path of the fluid through the plate will affect the uniformity of temperature in the plate, which influences the efficiency of the collector. If we use one tube, snaked back and forth across the surface, the plate temperature will be lowest at the inlet and highest at the outlet, increasing as the fluid moves from inlet to outlet. This will not be a uniform distribution—the price of being easy to construct. A better arrangement is to have a series of tubes in parallel, connected to inlet and outlet headers. This is more difficult to build but provides a more even temperature. Figure 12.15 shows these paths. The pressure drop through the panel

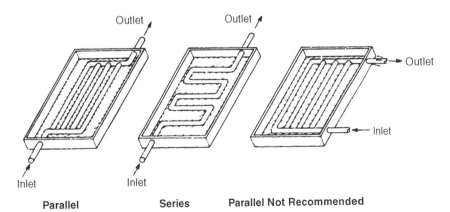

**Fig. 12.15. Paths of fluid through collectors. (*More Other Homes and Garbage*, Jim Leckie, Gill Masters, Harry Whitehouse, Lily Young. Copyright Sierra Club Books.)**

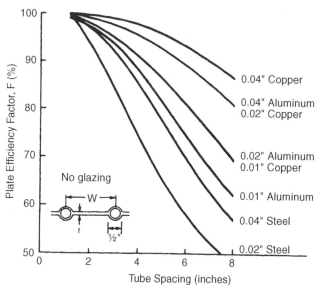

**Fig. 12.16. Collector efficiency versus tube spacing. (*More Other Homes and Garbage*, Jim Leckie, Gill Masters, Harry Whitehouse, Lily Young. Copyright Sierra Club Books.)**

determines the size of pump needed to move the fluid. A parallel path has less of a pressure drop than a series path and thus will require a smaller pump.

The spacing of the tubes across the panel also affects the efficiency. Wide spacing will give hot spots far from the tubes and cold sections near the tubes because the fluid in the tubes will remove heat from the section of the absorber in contact with the tubes. The thermal conductivity comes into play here. A copper plate will transfer heat much more readily across the absorber than steel, thus decreasing the temperature difference between different parts of the absorber. Figure 12.16 shows efficiency as a function of tube spacing for various materials. Of course, the closer the spacing, the more tubes must be used and, thus, the more expensive the collector.

A shiny copper absorber surface will reflect much of the incoming solar energy back into space. This will not help raise the temperature of the water. An obvious solution is to paint the surface with a flat black paint that will withstand high temperatures. The paint will typically absorb over 90% of the incoming energy. However, the reader will recall that if a body is warmer than its surroundings, it radiates heat to the surroundings. Flat black paint has an emittance almost equal to the absorptance at the wavelengths of concern, so an absorber painted with a flat black paint

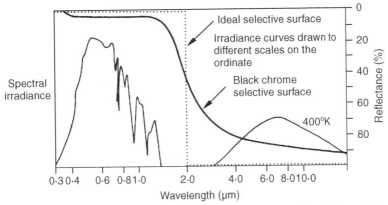

**Fig. 12.17. Spectral properties of several coatings. (*An Atlas of Renewable Energy Resources*, Julian Mustoe. Copyright John Wiley & Sons Limited. Reproduced with permission.)**

will emit nearly as much energy as it absorbed. If we can afford the cost, we should choose a coating that offers both high absorbtance and low emittance. Figure 12.17 shows the spectral properties of two coatings.

### Controlling Losses

Once we have done what we can to get the solar energy into the absorber panel, we need to keep it there to heat our water. Heat can leak out the top of the collector, through the glazing, or through the sides and bottom.

### Glazing

Collectors that operate at a plate temperature $<20°F$ above ambient require no glazing because radiation is low. When the temperature difference is $>20°F$ glazing is required to keep most of the heat in. The factors to be considered in choosing glazing material, in addition to being transparent to incoming radiation, include cost, appearance, resistance to impact and weather, low transmittance of reradiated wavelengths, and ease of working. Figure 12.18 shows plate efficiency factors with several glazing options.

Glass is the most common product for glazing. Table A.3 in the Appendix gives some of the properties of commercially available glass. Although glass is most common it has some drawbacks. It is difficult for an amateur to cut and handle, and even tempered glass has limited impact resistance. The other option is plastic. There are a variety of plastics on the market that give almost as much transmittance as glass while avoiding

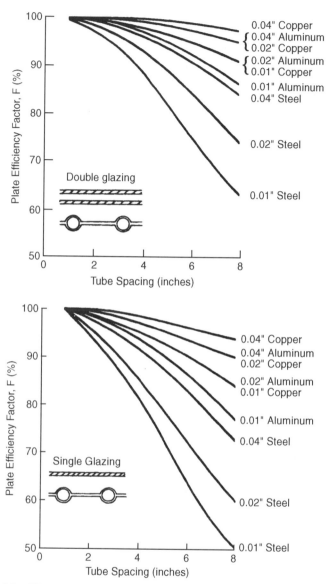

**Fig. 12.18. Plate efficiencies for glazing/spacing material options. (*An Atlas of Renewable Energy Resources*, Julian Mustoe. Copyright John Wiley & Sons Limited. Reproduced with permission.)**

some of the drawbacks. High operating temperatures require that the type of plastic be carefully chosen to avoid softening and warping. Figure 12.19 shows the transmission through some glazing options.

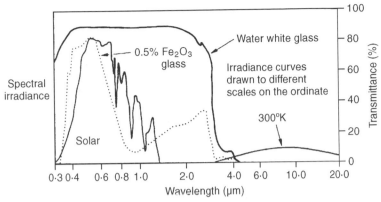

**Fig. 12.19. Spectral properties of glazings. (*More Other Homes and Garbage*, Jim Leckie, Gill Masters, Harry Whitehouse, Lily Young. Copyright Sierra Club Books.)**

### Insulation

Because the greatest heat loss is through the glazing, it is usually cost-effective to limit insulation to the equivalent of 3 in. of fiberglass on the back and 1 in. on the sides.

On a warm summer day, a noncirculating collector may reach temperatures >200°F. This implies that the insulation material must withstand this high heat without degradation. One option is fiberglass with a low binder content. Binder is added to fiberglass to give it shape; however, the binder tends to outgas at high temperatures. The result of outgassing is a cloudy coating on the glazing that reduces the efficiency of the system. The only foam insulation suitable for this application is isocyanurate, which is rated to 400°F and will work with painted absorbers.

### Housing

The collector housing has several functions. It provides mechanical support for the absorber plate and glazing and it protects the plate from the weather. Housings are normally constructed of steel, aluminum, or weather-resistant wood. The housing should be designed to give a water-tight seal between the glazing and the housing, water vapor between the glazing and the absorber will itself absorb radiation. The housing should be constructed in such a way that the glazing, the most fragile part of the collector, can be replaced if it is damaged. The weight of the housing is important because the overall weight of the system determines how strong the supports must be and what kind of equipment will be needed to lift and install the system.

### Collector Performance

We need an estimate of the performance of the system. This will tell us if it is worth the time and effort to install a flat plate water heater.

What we are trying to do is absorb incoming solar radiation and transfer it to a working fluid. The rate of heat transfer is the difference between the energy absorbed by the plate and the energy lost from the plate to the surroundings:

Rate of transfer $(Q_u)$ = energy absorbed $(Q_{abs})$ − energy lost $(Q_{loss})$

[12.8]

The energy absorbed is the product of the energy striking the surface times the transmission of the glazing times the absorbtance of the plate.

$$Q_{abs} = \text{insolation } (I) \cdot \text{area } (A) \cdot \text{transmittance } (t) \cdot \text{absorbance } (a)$$
$$Q_{abs} = I \cdot A \cdot t \cdot a$$

[12.9]

The losses to the surroundings are given by

$Q_{loss}$ = collector heat loss coefficient (U1) · area $(A)$
$\quad$ · difference in temperature between plate and ambient $(T_p - T_a)$

$$Q_{loss} = \text{U1} \cdot A \cdot (T_p - T_a)$$

[12.10]

If we combine the equations for energy absorbed and energy lost we get the following:

$$Q_u = A \cdot (I \cdot t \cdot a - \text{U1} \cdot (T_p - T_a))$$

[12.11]

The efficiency of the system is

$$\eta = \frac{\text{energy collected}}{\text{energy incident}} = \frac{Q_u}{I \cdot A}$$

[12.12]

or

$$\eta = t \cdot a - \frac{\text{U1}(T_p - T_a)}{I}$$

[12.13]

Figure 12.20 shows some typical efficiencies of various collector types. This figure indicates the following:

- The efficiency of all collectors decrease as the temperature difference between the outside air and the plate increases.
- There is no universally efficient collector.
- Glazing reduces efficiency at small temperature differences because it reduces incoming insolation.

For example, if we are heating water for a swimming pool and want a temperature of 80°F when the ambient temperature is 65°F, we are probably better off with an unglazed collector delivering water at 85°F rather than a sophisticated collector delivering water at 140°F. If our task is to supply domestic hot water at 140°F (a temperature difference of 70°F) then a glazed collector with a selective coating is required.

To design an effective collector we need to answer the following questions: What is the temperature difference under normal operating conditions? What size of collector must we install to meet our needs? The temperature difference will determine the type of collector. Figure 12.20 can be used as a selection guide. Determining the size of the collector is a more involved. So far, our efficiencies have been instantaneous. What we need to find is the performance over a day or several days: How much water can be heated during a day or during several days? We need to know how

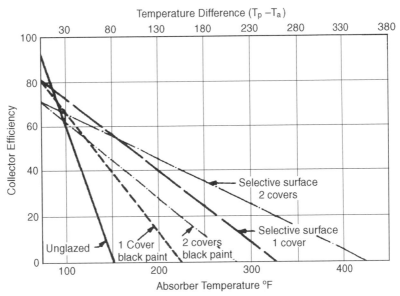

**Fig. 12.20. Collector efficiencies. (*More Other Homes and Garbage*, Jim Leckie, Gill Masters, Harry Whitehouse, Lily Young. Copyright Sierra Club Books.)**

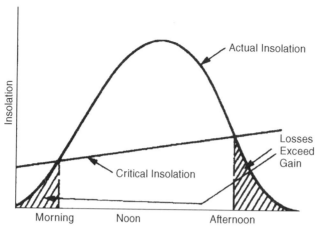

**Fig. 12.21. Insolation versus time. (*Energy Primer Solar, Water, Wind, and Biofuels*, Richard Merrill. Copyright Dell Publishing.)**

much heat will be collected. First, we must consider how many hours a day we will operate the collector. When energy absorbed is less than the energy lost it makes no sense to operate the collector; that is, we do not run a collector at night. Figure 12.21 shows insolation versus time and the minimum insolation ("critical insolation") needed for operation. Actual values for insolation versus time of day can be found in some of the handbooks listed in the References. Once the turn-on and turn-off times have been estimated we need to determine the total expected insolation between those times. Again, actual values can be found in weather records in handbooks.

Massaging the equations given earlier in the chapter (using the same approach as in Eq. [12.7]), we derive

$$\text{Area} = \frac{\text{energy required}}{(\text{insolation efficiency})}$$

This equation should make intuitive sense — area equals the total energy required divided by the energy gained per unit area.

We present an example to clarify these ideas. Suppose we have a family of four requiring 20 gallons of hot water (140°F) per person per day. The ambient temperature is 50°F. Recall that a Btu is the amount of heat required to raise the temperature of 1 lb of water 1°F. A gallon of water weighs 8.34 lbs. The energy required to heat the water is

$$Q = 20\,\frac{\text{gallons}}{\text{person} \cdot \text{day}} \cdot 4\,\text{people} \cdot 8.34\,\frac{\text{lbs}}{\text{gallon}} \cdot (140 - 50)°\text{F} = 60{,}048\,\frac{\text{Btu}}{\text{day}}$$

From a handbook we determine the insolation versus time for January 21 at 32° north latitude; this day is chosen because it is a difficult example.

| Time | Insolation (Btu/hr/ft$^2$) |
|------|----------------------------|
| 7 AM | 0 |
| 8 | 123 |
| 9 | 212 |
| 10 | 274 |
| 11 | 312 |
| 12 noon | 324 |
| 1 PM | 312 |
| 2 | 274 |
| 3 | 212 |
| 4 | 123 |
| 5 | 0 |
| Total | 2166 |

With a temperature difference of 90°F we would probably select a collector with a selective surface and one layer of glazing (see Fig. 12.20). From Fig. 12.20 we determine an efficiency of about 55%. Therefore, the area is

$$A = 60,048 \text{ Btu/day}/(2166 \text{ Btu/ft}^2 \text{ day} \times 0.55 \text{ (efficiency)}) = 50.4 \text{ ft}^2$$

A collector with 50.4 ft$^2$ of absorber area would provide the energy required to heat 80 gallons of water from 50° to 140°F.

### Collector Plumbing

Now that we've devised a way to capture heat from the sun, we need to move the heat somewhere useful. Because the demand for hot water does not usually coincide with peak insolation, some storage is required. A method of preventing freezing is also necessary.

The simplest approach to these problems is a thermosiphoning system. Thermosiphoning applies the principle that warm fluid is less dense, and therefore lighter, than cold fluid. Warm fluid will rise to the top of the system and cold fluid will sink to the bottom. We can take advantage of this rise of warm fluid by placing our storage tank above the collector. Figure 12.22 shows a thermosiphoning arrangement. Cold water is let into the bottom of the storage tank. If the collector is warm, the coldest water, being densest and therefore heaviest, will drop to the lowest point in the system—the bottom of the collector. As the cold water goes into the collector it forces the warmed water that had been in the collector into the

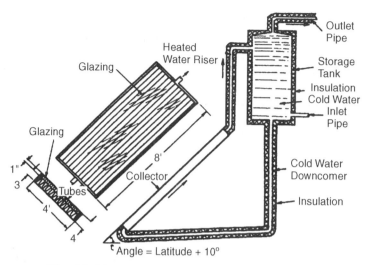

**Fig. 12.22. Thermosiphoning hot water system.**

upper part of the storage tank. In fact, the flow is continuous, even if no cold water is added or hot water taken out. As the water in the storage tanks cools it becomes heavier and flows down into the collector, displacing the lighter warm water, which goes to the storage tank.

When a hot water tap is opened hot water flows out of the storage tank. Cold water then can flow into the storage tank. Thermosiphoning will continue in the collector and storage tank, whether or not there is any demand, until the water in the storage tank is as warm as that in the collector.

This system is simple, requires no pump, and works well if properly designed and maintained. However, there are problems related to tank placement, to a possible spell of cloudy weather, and to freezing. We consider tank placement first.

If the collector is roof mounted the storage tank must be placed in the attic or rafters. The tank, when full, will weigh about 700 lbs and will require special structural support. Should it leak the damage may be considerable.

If one adds a small pump to raise the cold water to the collector, one is freed from the storage tank placement constraints of the thermosiphoning system: One does not have to depend on natural siphoning, which requires that the tank be positioned above the collector to move the water. Pumping the water allows a more conventional arrangement, with the collector on the roof and the storage tank in the basement or utility room.

We consider how to avoid the problem of a spell of cloudy weather. An auxiliary heater is our concession to the reality that our collector cannot always provide hot water. Cloudy days require that we have some sort of backup heating. The alternative to an auxiliary heater would be an impractically large collector and storage tank. The auxiliary heater is more economical and also makes the system more reliable by providing a backup heater if something goes wrong in the collector.

The system shown in Fig. 12.22 will be damaged if the water in the collector freezes. One solution is to drain the collector when there is any danger of freezing — a significant nuisance. An alternative to draining the system each time there is a threat of freezing is to circulate antifreeze rather than water. Obviously, one does not want to shower in antifreeze. To avoid that discomfort the antifreeze is circulated through a heat exchanger in the storage tank. A heat exchanger is a length of tubing (usually copper), sometimes with fins, that transfers heat from one fluid to another; that is, it provides a path for efficient heat transfer but prevents mass transfer. The hot antifreeze from the collector flows through the heat exchanger, which warms the water in the storage tank.

The variety of systems and components available for hot water heating allows users a great deal of flexibility in choosing the system that best meets their lifestyle, the amount of money they want to spend, and the amount of hot water they require.

## HOUSE HEATING

It was mentioned that collectors are not usually used for heating a house. If, for some reason, one did want to use a collector for space heating then one could calculate the required area as done in Eq. [12.7] but change the efficiency factor from 0.20 because collectors are more efficient than direct gain absorbers. Efficiencies could be estimated from Fig. 12.20.

## PROSPECTS FOR SOLAR HEATING

Solar heating seems to be a technology ready to become widespread. If the cost of alternative systems, usually fuel oil costs in this case, rise appreciably, solar heating will become economical. Currently, in the Northeast solar hot water should be considered seriously when new dwellings are planned. Solar heating is certainly a technology that a user can understand, maintain, and even, at least in part, design and construct.

# REFERENCES

1. Johnson, T. E. (1981). *Solar Architecture — The Direct Gain Approach*. McGraw-Hill, New York.

   *Readable discussion with good pictures; includes data on solar radiation and temperature for about 200 cities and R values for many building materials.*

2. Leckie, J., Master, G., Whitehouse, H., and Young, L. (1981). *More Other Homes and Garbage*. Sierra Club Books, San Francisco.

   *Through treatment of both theory and construction; much data on both climate and materials; discusses the same things as in this chapter but in much more depth.*

3. Shapiro, A. M. (1985). *The Homeowners Complete Handbook for Add-on Solar Greenhouses and Sunspaces*. Rodale Press, Emmaus, PA.

   *Readable, complete discussion of both greenhouses and how to grow plants in them. The author has had much practical experience and explains things well; includes costs and lists of suppliers.*

4. Department of Housing and Urban Development (1982). *Passive Solar Homes*. Facts on File, New York.

   *Wide collection of actual house plans, with some text and a sample calculation of heat loss; good ideas for someone planning to build a house.*

# PROBLEMS

1. Are Figs. 12.5 and 12.6 consistent in terms of human vision; that is, does the human eye see the wavelengths that the sun produces most strongly?

2. How much of the sun's energy needs to be collected to replace other energy sources in a typical house? Assume that each day the house uses 15 kW-hr of electricity, 1000 ft$^3$ of natural gas, and 7 gallons of fuel oil. A cubic foot of natural gas contains 1 Btu and a gallon of fuel oil 140,000 Btus. If the sun's energy is available for 8 hours how large an area is needed to collect the required energy?

3. Repeat Problem 2 using estimates from a house you know about. Would the roof be large enough to collect this amount of energy?

4. Estimate the thermal resistance of the wall in Fig. 12.2 if the insulation were 7 in. thick.

5. Estimate the energy needed to keep a 10 × 20 m swimming pool at 25°C if the temperature of the moving air is 20°C.

6. Estimate the heat energy loss in the house in Fig. 12.3 if the walls had no insulation — they consisted of plaster directly on wood.

7. Estimate the energy loss in the house in Fig. 12.3 if double-glass windows are used.

8. A cord of wood contains about 20,000,000 Btus and a woodstove is about 50% efficient (about half the heat goes up the chimney). How many cords a year would the house in Fig. 12.3 need?

9. Explain why storage is needed in a direct-gain heating system.

10. Redesign a house you know about to use direct gain for heating.

11. A rolled up shade is shown in Fig. 12.11, the Trombe wall heated house. Explain why this shade should be dropped at night.

12. A swimming pool loses heat by convection and also by evaporation of water, by conduction, and by radiation. It gains heat directly. Assume a pool needs 600,000 Btus/day in May and September, how large a collector would you build?

13. (Design project) Design a solar greenhouse as an addition to a house you know about. Aim for about 50,000 Btu/day.

14. (Design project) Design a vacation house on a site you know about, paying particular attention to landscaping, orientation of house, and location of rooms.

15. (Design project) Why might an underground house be energy efficient? Design such a house.

16. (Design project) Design a solar collector that produces 100,000 Btu/day. Specify materials and all construction details. Choose a geographic location, perhaps your own home.

# APPENDIX

## Table A.1. Heat Transfer Coefficients

| Material | Conductive Heat Transfer Coefficient $((Btu \times in)/(hr \times sq\ ft \times °F))$ |
|---|---|
| Aluminum | 1400 |
| Brass | 715 |
| Brick | 5 |
| Cellulose | 1.66 |
| Clay, dry | 4 |
| Concrete | 12 |
| Cotton | 0.39 |
| Glass wool | 0.27 |
| Glass | 5.5 |
| Granite | 15.4 |
| Gyspum | 3 |
| Ice | 15.6 |
| Iron | 326 |
| Limestone | 10.8 |
| Paper | 0.9 |
| Plaster, Cement | 8 |
| Sand | 2.28 |
| Soil | 12 |
| Water | 4.1 |
| Wood | 0.9 |
| Wool | 0.264 |

## Table A.2. Convective Heat Transfer

| Wind Speed (MPH) | Convective Heat Transfer Coefficient (Btu/(hr × sq ft × °F)) |
|---|---|
| 0 | 1 |
| 2 | 1.6 |
| 4 | 2.2 |
| 6 | 2.8 |
| 8 | 3.4 |
| 10 | 4 |
| 12 | 4.6 |
| 14 | 5.2 |
| 16 | 5.8 |

## Table A.3. Properties of Glazing Materials

| Material | Solar Transmittance (%) | Reflectance (%) | Vertical Shading Coefficient |
|---|---|---|---|
| Acrylic | 85 | | 0.98 |
| Polycarbonate | 90 | | |
| Clear Glass | 83 | 8 | 0.99 Single |
| Starphire Glass | 90 | 8 | 1.05 Single |
| Solex Glass | 61 | 6 | 0.81 Single |
| Clear Glass | 69 | 13 | 0.88 Double |
| Solex Glass | 51 | 9 | 0.69 Double |

The shading coefficient is the ratio of the total amount of solar energy that passes through a glass relative to 1/8-inch-thick clear glass.

# *13*

· · · · · · · · · ·

# APPROPRIATE TECHNOLOGY IN THE THIRD WORLD

· · · · · · · · · · · · · · · · · · ·

The first part of the book dealt with various technologies themselves — the hardware. We discussed how to design things, how much they produce, and how much money, time, and material is required. Knowing these things is essential for deciding if Appropriate Technology meets the needs of the Third World. In the second part we consider less tangible factors — how does choice of a technology affect the way people live, how they interact, and what they think of themselves? An ancient saying is that people do not live by bread alone. We want to think about what else they live by. We want to decide if Appropriate Technology has something to offer for people's personal, social, political, and cultural lives.

Two practical reasons exist for considering these factors. First, if we are responsible for technological change, we must anticipate the influence this change will have, beyond the hardware. A medieval historian, Lynn White [21], has shown how the invention of the stirrup, which made it possible for knights to fight on horseback, led to the age of chivalry. Television has changed politics in the United States. Before television, national candidates could not hope to interact with most voters. The interaction was through the party. Television, though, created a direct channel between candidates and voters, so the party organization has lost importance. The high cost of television ads has also changed politics. These examples show that the unexpected and unintended consequences of a technological change can be great.

The other reason for considering intangible factors is implementation. Success in making change requires commitment from the people involved. To get this commitment it is necessary to understand the attitudes and concerns of those affected and these attitudes and concerns usually go beyond material costs and benefits. Schemes in the Third World to prevent deforestation which made economic sense have failed because wood collectors were neither consulted nor recruited. In Chapter 11, we demonstrated how people's attitudes, which had nothing to do with making or losing money in this case, determined the effectiveness of pit latrines. In summary, to make valid technological decisions, we must understand the technology and also the broader consequences of its implementation. The implications of our decisions will be far reaching and considering the human factor is essential in making change.

The central question in this chapter is whether Appropriate Technology really has anything worthwhile to offer the Third World [18]. We start by examining the needs of the Third World — both material and attitudinal. We note that any effective strategy for meeting those needs must not solve immediate problems by creating long-term ones, for example, a farming program which yields much food for a few years but irretrievably ruins the environment is not desired. An example of the use of an inappropriate technology in the Third World, space-age photovoltaics in the Marshall Islands, is described. The argument is made that Appropriate Technology can give the Third World food, shelter, energy, and other necessities without forcing dependence on other nations and without destruction of the way people live. Some examples of successful Appropriate Technology projects are given. Finally, two problems with Appropriate Technology are examined — that it does not really meet needs and that it does not lead to the advancement of the nation employing it.

# NEEDS IN THE THIRD WORLD

In order to understand what Appropriate Technology can offer the Third World, we consider needs in Third World countries. The Third World is not one homogeneous place and needs differ from locality to locality. Our discussion can, though, suggest the requirements for people's well-being. People in the Third World are no different than people anywhere else: They need food, clothing, shelter, and health care. The technology chosen must not only meet these needs now but also must be sustainable; that is, it must be able to continue to meet needs. In particular, a sustainable strategy is one that preserves the environment — agricultural land, forests, rivers, and wildlife. Beyond material needs are the need for attitudes that promote self-reliance and improvement. People also have a need to be part of, and enjoy, a common culture, to feel part of a group and to appreciate sports, art, and music. Culture is a main subject of Chapter 15, in which it is argued that Appropriate Technology acts to support the cultural heritage of Third World nations.

## *Material Needs*

We first examine the necessities of a nation as a whole. Foreign exchange is an essential for any nation because it is not possible for a country to produce all that it uses. (Only recently has the United States seemed to realize its dependence on international commerce.) Less industrialized countries need foreign exchange to buy manufactured products. In many cases, a major use for foreign exchange is oil — chiefly gasoline and diesel fuel. While some imported products may seem frivolous (VCRs and stereo systems), others are certainly not (e.g., medicine). A plan for a Third World nation which does not include exports earning foreign exchange impossibly limits life there.

An entirely different aspect of meeting material needs is where they are met. Most of the people in the Third World live in villages, although many people have left the villages and settled in cities, often in makeshift housing on the outskirts of cities (e.g., Mexico City). The cities often do not have the infrastructure — water, sewage systems, hospitals, schools, and so

**Fig. 13.1. Wedding preparations in an African village.**

forth — to cope with the number of people moving in. Therefore, an approach to meeting material needs is desired which makes village life an attractive alternative to the city. Besides basic needs such as food and shelter, a need exists in the village for amenities, a higher standard of living. Villages lack stimulation and entertainment so the availability of television and recorded music can make village life significantly more attractive. Stereos and VCRs may not be frivolous needs at all. Figure 13.1 shows a village wedding feast, but such festive occasions are too expensive and time-consuming to occur often.

## Sustainable Development

It was stated previously that the program for meeting material needs must be "sustainable." What does this mean? The economies of most Third World countries are based on their environmental resources — soils, forests, fisheries, wildlife, and parks. These resources are being rapidly depleted. The loss of forest in the Amazon basin and the enlargement of the Sahara desert are prime examples of this depletion. Similar depletion of resources occurs in much of the Third World. A sustainable strategy would be one that maintained the resources while using them, just as one can use the interest from a bank account without withdrawing the principal.

The specifics of sustainable development are straightforward. In agriculture, the quality of the soil is maintained or enhanced. In forestry, each tree taken is replaced. In manufacturing, both material and energy are used efficiently, with much recycling of raw materials and use of renewable energy sources. Air and water pollution are minimized.

Sustainable development usually refers to natural resources but the idea is also useful in another context. Development is only sustainable if it will survive after the originators leave. Development projects are often started by people from Europe and the United States, so-called "expatriates," visiting Third World countries. The question is what happens to the project when the expatriates go home.

## *Environmental Concerns in the Third World*

In many countries throughout the world, environmental issues balance immediate needs with future possibilities. In the Third World, though, immediate needs often seem more pressing, especially if the country's foreign debt is large. If the requirements for foreign exchange for an oil-exporting country are imperative then it will seem necessary to sacrifice the environment to the oil industry, which is why areas of Mexico are some of the most polluted on earth. Exhortations from the United States to the Third World to protect the environment, even if industrialization slows, are not always well received. Third World leaders genuinely question if they can afford to protect their environment. The government of China defends its damming of the Yangtze River by citing the need for economic development.

An analogous trade-off exists between an individual's need for material things and fear of industrial hazards. A U.S. journalist, Fergus Bordewich, went to Bhopal a year after the disaster expecting to find hostility toward the United States and toward industrialization in general [4]. Little of either was found, however. A worker, crippled in the disaster, said he would return to work tomorrow in the same plant if it reopened. The Indian government continues to court foreign industry and has made only modest environmental restrictions. The journalist concludes that the "security of a good wage and the dignity of a modern job" are the most powerful forces at work. Such is also true in the United States, in which workers often choose a job that is hazardous because of pollutants over a job at lesser pay. The Occupational Safety and Health Act was implemented to protect just these workers. A Brazilian national leader called poverty the worst pollutant of all. The Third World needs a development strategy that does not force trade-offs between having necessities now and having environmental resources later.

## *Encouragement of an Attitude of Self-Reliance*

Certain attitudes, including self-reliance and self-confidence, are essential if people are to continue to improve their lives. President Kenneth Kaunda of Zambia, on the tenth anniversary of Zambia's independence, wrote with evident approval of "the change in Zambia's people—proud and self-confident instead of second class citizens" [9]. We describe some attitudes that appear to be essential if a society is to prosper, materially and culturally.

One such attitude is that local people, on their own, can make life better. The harm of an attitude of resigned hopelessness is obvious. The harm of an attitude of dependence—expecting someone else, a foreign government or aid organization, to solve the long-term problem—is just as pernicious. Not all aid programs, though, have recognized the need for eliminating continued dependence, even on such mundane items as spare parts or technicians. Development programs based on technologies that local people feel they can never master will not produce an attitude of confidence.

Another attitude that is essential for long-term survival is responsibility—the attitude that the success of a venture depends on what individuals actually do. A scheme for improving village life will accomplish little if, in the long run, the people remaining in the village do not feel responsible for it. Irrigation channels will clog. Farm machinery will break down and rust at the edge of a field. Health centers will be unused unless the people in charge believe it is possible and important to keep them functioning. Two components of responsibility are beliefs that one can accomplish a goal and that the goal will improve life. It is essential to ensure that a project provides what is necessary—training, development of confidence, tools, emergency assistance, and so forth—so the person who will run it when the expatriate team leaves believes herself/himself to be responsible.

A third valuable attitude is respect for the existing national accomplishments of a nation. A harmful attitude is that progress comes to a person or to the nation only by abandoning village life. Another such harmful attitude is that whatever is done in the United States or Europe—in popular music, education, agriculture, and so forth—is best. This latter belief originated at least in part because the Europeans, and to a lesser extent U.S. citizens, had been colonial rulers Present-day Third World leaders speak of the necessity for "decolonization of the mind"—eliminating the belief that copying former colonial rulers is the proper way [7].

An approach used to develop an attitude of self-reliance is that of promoting community participation—assisting local people to organize and systematize their knowledge, enabling local groups to take action, connecting them with outside agencies. Examples include bringing a rural community in Kenya together to produce a community action plan, partly to build a water supply and partly to improve village life in other ways [16], and involving communities near the boundary of a national park in Zambia in game management and giving them a share of the income [19]. The "Campfire" program in Zimbabwe accomplishes the same goals by giving villages on the borders of national parks control over money raised from hunting licenses for their areas [13].

We have been discussing the needs of the Third World because we want to determine how well Appropriate Technology meets those needs. To make the discussion more concrete we will describe a situation in which solar cells, which certainly seem to be an Appropriate Technology, were introduced into a Third World community.

## AN EXAMPLE OF A NEW TECHNOLOGY IN THE THIRD WORLD

This example details the installation of a photovoltaic system on one of the Marshall Islands. The installation was evaluated by Ralph Beckman, a respected architect and designer who had previously worked on photovoltaic systems in Botswana as well as several major solar projects in the United States. A videotape of Mr. Beckman describing the Marshall Island project is available [3].

The Marshall Islands is a trust territory of the United States administered by the Department of the Interior, which is also responsible for Indian Affairs. The island in which we are interested is called Utrek (Fig. 13.2). A well-known Marshall Island is Bikini, the site of atomic bomb tests in 1954. (The name came into general use for a bathing suit after that test.) Another well-known Marshall Island is Kwajalein, currently used by the U.S. armed forces as the target for long-range missile testing.

During the 1954 tests the wind shifted unexpectedly and the people of Utrek, several hundred miles from Bikini, received radioactive fallout. At the time they were told to be careful with their drinking water and to slaughter, but not eat, their livestock (chickens and pigs). Each was given $49 to "cover the loss of these animals." Thirty years later numerous cases of thyroid cancer appeared among those islanders who had been present during the fallout. A lawyer in Washington took up their case and the Islanders were provided medical care and an award, for the entire island, of about $100,000. Islanders have minimal use for cash so a project of general use was deemed appropriate by the trust administrators. Hughes Corporation offered a surplus photovoltaic (PV) system designed for spacecraft

**Fig. 13.2. Utrek Atoll.**

which was purchased for the $100,000. NASA supported this venture to show how its research and development activities can have useful commercial spin-offs.

As part of the deal, Hughes agreed to train a local person to maintain the system. Fortunately, a person from the island had been a mechanic for the Navy. His name was Lembert Debru. It was clear from Lembert's record and his discussions with Hughes employees that he was able and responsible. Figure 13.3 shows Lembert with the batteries of the installed system. Lembert was sent to Long Beach, California, for a week of training and orientation. He was given a full set of instruction manuals. On the last day of his visit it was learned that Lembert did not read English — nobody had thought to ask. However, the system was installed as an experiment. A year later Mr. Beckman was asked to go to Utrek and be part of a team evaluating its success. He found the PV devices to be working well but the batteries were failing — possibly because conditions on a hot, humid island are so different from those in space. Because of the condition of the batteries the system was close to total failure.

Mr. Beckman also found that the system was being used differently than expected. When the system was installed it was decided to divide its output evenly among the village homes. This division meant that one fluorescent light fixture and one night-light were installed in each house. A typical home is shown in Fig. 13.4. Social custom in the Marshall Islands, where it is hot and humid, is that people spend their evenings outside the houses. Therefore, some people had moved their fluorescent fixtures

**Fig. 13.3. Lembert with the batteries for the PV system.**

**Fig. 13.4. Home on Utrek.**

outside using wires and homemade supports — the islanders are good at adapting designs that can be adapted. Others simply ignored the whole thing. The night-light did seem to be used though. Small shrines were part of many homes and the night-light was placed in front of these — replacing candles or kerosene lamps.

Mr. Beckman, not surprisingly, was concerned that all the islanders' money had been spent for a system that a year later was on the brink of collapse and that, at best, would not improve their lives significantly. He appealed to NASA for help but the office responsible for the project had been eliminated in a reorganization. After a good deal of trouble, he was able to get support to the island and make improvements on the battery system.

What can we learn from this example? Mr. Beckman contrasts this solar power system with the traditional sailing craft used by islanders (Fig. 13.5). These boats are beautifully designed, built of local wood by hand. Islanders sail them beyond the sight of land, navigating skillfully by the stars and by signs from the sea and sky invisible to nonislanders. Legends are written about such voyages. Owners know how to repair their boats and the materials are locally available. The boat serves a real need — fishing and inter-island transportation. The boat does not consume foreign exchange. The boat uplifts the spirit — it is a graceful piece of work, capable of doing heroic things. It is part of the culture and enlarges that culture. The PV system is none of these. Few repairs can be done locally because neither parts nor expertise are available. People cannot feel confidence in it because it is not understandable. The system does not meet an important need for the villagers. It is wasteful of resources — money in this case — and it does not raise the spirit.

**Fig. 13.5. Traditional sailing craft.**

# WHAT APPROPRIATE TECHNOLOGY CAN OFFER THE THIRD WORLD

A lesson from Mr. Beckman's story is that a new technology, space-age photovoltaics, did not meet the needs of a community. It is probably obvious that this technology did not fit our definition of appropriate. It was not understandable nor controllable by the user. It depended on foreign components, the batteries. It is capital rather than labor-intensive. Even the decision to make one large system for the whole community rather than put individual panels on the roofs of each home is incompatible with the decentralization aspect of Appropriate Technology. The traditional sailing craft fits our definition of Appropriate Technology and meets local needs.

The discussion that follows, like Appropriate Technology in general, consists mostly of examples which succeeded in one place rather than grand programs which can be replicated everywhere. If Appropriate Technology is more effective than complex technology, the reader might ask why Appropriate Technology is not used exclusively. One answer is that it does not accomplish all goals. It offers certain, but not all, benefits. Also, Appropriate Technology is inherently small scale and does not easily attract attention—its benefits can be overlooked even though the many small improvements it produces add up to more than that of a few large-scale, highly visible successes.

Appropriate Technology offers (i) goods and services, on a sustained basis; (ii) vehicle for independence; and (iii) gradual transition. Goods and services, of course, meet the need described previously for material necessities. Both independence and a gradual transition encourage the attitudes effective in supporting confidence and responsibility. Both also foster preservation of the national culture.

## Goods and Services

In this section we consider examples of when Appropriate Technology has been the best way to provide goods and services. The argument here is the same as Schumacher's—that in some situations Appropriate Technology is the only feasible way to meet the material needs of the Third World. The alternative, high-technology, requires an unrealistically large amount of capital and often does not meet real needs (as the Marshall Islands example shows). Of course, examples exist in which high technology is successful in improving the condition of people in the Third World.

Small-scale agriculture has certainly become tremendously more productive since the introduction of new kinds of seeds, with irrigation and chemical fertilizers. Small farm machinery, such as hand-operated tillers (Fig. 9.1), has increased output even more. Improvements in technology have made the lives of individual farmers better without causing them to lose their autonomy and existing lifestyle. These farming improvements do bring some dangers. For example, widespread drawing of water from deep wells, sometimes required for irrigation, can lower the water table, leading to desertification. Dependence on imported fertilizer is risky if the nation becomes short of foreign exchange. Individual farmers are at risk if they must borrow money to pay for fertilizer and plan to use the proceeds from the harvest for repayment. On balance, though, use of these new seeds — the Green Revolution — has made small-scale farming in the Third World profitable and given farmers a surplus that can be used to make life better. Critchfield [6] describes how much village conditions have improved in Indonesia since the high-yielding seeds were introduced: Farmers have brick houses with tile floors and glass windows, their own well, and so forth. Highly mechanized farming would probably not have been profitable in this hilly terrain and certainly would have required a major investment. Furthermore, small-scale farming is easier on the environment than large-scale farming and is much more likely to be sustainable. Small-scale agriculture is Appropriate Technology and has improved farmers' lives.

In Zambia, outside the capital city of Lusaka a self-help housing scheme was instituted as an alternative to shantytowns or high-rise apartments. The local government divided the area into reasonably sized house lots and on each lot built a cubicle with running water and a sewage connection. The cubicle could become a toilet and a shower with a sink on the outside. The purchasers of the lot were given house plans. They then built their own houses, using traditional methods if they chose. Simple presses to make bricks, designed by a European aid organization, were available so substantial brick houses could be built at a cost only slightly greater than that of traditional houses. Many people did choose to build from brick. The community is now fully occupied and well regarded.

Another successful self-help scheme was the provision of clean water in villages near the mountains in Malawi, an example used in Chapter 11. Engineers from the Ministry of Rural Development located sites along mountain streams suitable for collecting water. Purity of the water and dependability during the dry season, as well as accessibility, were the main criterion in choosing sites. The engineers also explained how the system would work and how to lay the pipe. The government supplied polyvinylchoride pipe. The people in the villages chose the distribution points, where the pipes would go, and actually did the construction. The project cost less and was better built than if a government crew had installed it. Maintenance work could be done by the people using the water, although actually very little maintenance has been required.

Craft workers making traditional umbrellas as well as wood carvers and lacquer work artisans (Fig. 13.6) have prospered in Thailand. These people live and work in villages — selling some of their products directly to visitors but usually dealing with exporters and wholesalers. This is not primitive technology. It has become increasingly effective as improvements are recognized and incorporated. The products lend themselves to small-scale production and are in demand as an export good. The work appears to be more challenging and working conditions, in open sheds, more attractive than a likely alternative — bonding semiconductor chips in a large factory. Rugmakers in Turkey and wood carvers in Africa are other examples of craft workers that compete effectively with higher technology manufacturing, prospering commercially while performing creative work in healthy conditions. This form of production is energy efficient and wastes little material. It is more likely to be sustainable than machine-based manufacturing.

**Fig. 13.6. Umbrella maker in Thailand.**

Other craft workers may simply meet a local need more economically than do imported products. An example is metal workers who fashion buckets and garden tools from scrap metal. The work requires skill and training and the product is just as serviceable and much cheaper than the alternative. The buckets shown in Fig. 13.7 were made in Africa from scrap pieces of sheet metal roofing. Clay pots and dishes are another example of products which can be made as well by local crafts people as by machines. Village women in most of Africa are skilled at making cooking pots and in several places have transferred those skills to making plates, cups, and dishes, some of which are exported. In Botswana a wide range of products

**Fig. 13.7. Products of small-scale technology in an African market.**

have been made by small-scale industries, including farm implements, textiles, jewelry, furniture, and leather goods. The organizations making these products seem uniformly to have been financially stable and most have made the transition from expatriate management to local control [8]. A logo for one of these organizations is shown in the frontispiece to Chapter 17.

Each of these examples — agriculture, housing, water supplies, health care, and manufacturing — shows that a small-scale approach can offer goods and services of good quality at low prices. Each is at least as gentle to the environment as the high-technology approach and each promotes satisfying work. Appropriate Technology has been successful in these cases.

## *Vehicle for Independence*

Independence means many things. To a nation it means independence from foreign experts and imported goods. Imported technology does not always provide this independence. The experience of a large locally owned agricultural company in Nigeria may be instructive in considering imported technology and independence. The company's management recognized a developing market for eggs as more city people earned cash. Few local people had been trained in large-scale agriculture so cooperative agreements were signed with a European company. The Nigerian company built a large chicken "factory" in which chickens were kept in single cages and fed a special formula to produce eggs efficiently — a standard design in Europe. The feed formula had to be imported and thus was paid for by foreign exchange. After a few years, the price of oil, Nigeria's main export, on the world market fell so foreign exchange became limited. Feed could not be imported and so the operation had to shut down. The foreign partners were not strongly committed to the venture; it was just another investment to them so they did not push hard to keep it going. If the project had been planned and managed by local people, other sources of feed or other ways of keeping the project going might have been found. The foreign partners did not put a high priority on training local people — why should they? The project did not create independence for the people concerned. Of course, not all joint projects like this one fail but the risk exists.

Independence on a individual scale means an opportunity to be a responsible official in a company or even to have a job at all. If all the important decisions are made somewhere else in an organization, especially in a different country, then local people are not really independent. In a high-technology industry, a plant in a Third World country will almost surely be part of a multinational company and be a plant with a narrowly focused mission so its managers will have little autonomy. Furthermore, if it is difficult for a Third World person to gain some independence within a multinational company, it is even more difficult for her or him to be an owner of a high-technology company or an entrepreneur. The suppliers, the expertise, and the markets will not be present locally, so it will be very difficult to compete. Using an Appropriate Technology may give an entrepreneur more control over her or his destiny than would other technologies.

Appropriate Technology can be an effective way for a nation to become independent because it can be done with local people and local resources. If the technology is based on that existing in the country, training can be quicker—the transition from traditional farming to use of high-yield seeds in small gardens is less than that to high-technology farming. In the United States small businesses have created many jobs and it is believed that promoting small businesses is a more efficient way of creating jobs than supporting larger companies. The same seems to be true in the Third World. The United States government's policy is to work with organizations in the Third World which assist entrepreneurs. Appropriate Technology is well matched to the needs of entrepreneurs because these entrepreneurs are likely to have both the required knowledge and the raw materials. The March 1989 issue of *Appropriate Technology* [1] focuses on developing entrepreneurs.

It should be mentioned that the attitudes that an entrepreneur needs—self-reliance, resourcefulness, and self-confidence—are the same attitudes that were mentioned earlier as important for the sustained success of a nation. Perhaps that is why the small business assistance program is so highly regarded by development experts in the U. S. State Department. One project funded by the U.S. Agency for International Development in Ecuador made loans to 566 microindustries and nearly 2000 market vendors in less than a year, creating 2200 jobs and bringing "most importantly, an upsurge of hope for the future" [2]. The World Bank also supports small business assistance programs [14].

Furthermore, small-scale hand manufacturing can be a way into an industry, as it has been in the Far East where many electronics companies initially performed manual assembly of circuit boards and calculators. Now some of these companies are performing automated assembly of complex equipment. People who started by bending sheet metal in Kenya now have machine shops [10]. The technology that is appropriate in a situation depends on the experience of those involved; as more experience is gained, different technologies become appropriate. Managers and engineers are in a position to innovate if they are in control of what they are doing, as the traditional boat builder is in the Marshall Islands. People who are in control of their technology will be less likely to blindly follow foreign experts: They will have been decolonized. The Appropriate Technology route can lead a country away from dependence on other countries.

## *Gradual Transition*

As open land in Kenya and the Sudan is filled with farms, the nomadic cattle herders currently living there will be forced to settle in one place. How can the strengths of their culture, based on sharing and cooperation, be preserved despite the drastic change in their lives? How can the transition to a new way of life be made as easy as possible? Although it might make economic sense to set up a large plantation or even a factory to provide employment, small-scale farming and manufacturing would be more consistent with their previous lifestyle and thus will preserve better the traditional strengths of the culture. Industries which promote group work rather than competition between individuals are called for here, both to keep the society intact and to ease the adjustment to new ways of earning a living. Appropriate Technology tends to be more in line, in Third World countries, with traditional lifestyles than are Western technologies and thus its introduction tends to be gentler on the culture. This closeness also makes learning the new jobs quick and comfortable.

Appropriate Technology, because it does not create a big change, can avoid the creation of a two-tiered society — a highly trained elite supported by essentially untrained workers. Modern technology tends to use a few highly skilled workers; a factory run by robots is an extreme example. The main function of humans in such factories is programming the robots and responding to emergencies. When sophisticated, automated factories have been set up in a Third World country, often the result is that a small number of people are given advanced training, usually overseas, and nobody else is trained at all. A less sophisticated technology would result in the training of many more people, partly because the technology is similar to what had existed previously so training of many workers is practical and partly because more workers are employed. Furthermore, if an automated factory closes or a project is completed, highly trained specialists may not be able to find another job because similar work may not exist in the country.

## SUMMARY

The previous examples in this chapter show that Appropriate Technology can efficiently supply food, shelter, and jobs. The examples also show that Appropriate Technology can be less damaging to the environment than other technological choices. Appropriate Technology encourages people to be self-reliant and independent, the attitudes that seem to be essential for a sustaining and improving community. Because the work people perform after Appropriate Technology is introduced is not greatly different from what was done before, there is less disruption to the society. It does appear Appropriate Technology has something to offer the Third World.

## PROBLEMS WITH THE USE OF APPROPRIATE TECHNOLOGY IN THE THIRD WORLD

An article by Richard Critchfield is titled "Small is beautiful, but it's small" [5]. He suggests two concerns. First, that Appropriate Technology will not produce enough food or shelter or whatever is needed. Can it really meet people's needs? His second is whether Appropriate Technology leads to advancement of the nation employing it. Both of these concerns represent valid problems.

Appropriate Technology is not the most effective way to do everything; it cannot meet all needs. International telephone equipment and signals in one country must be consistent with those of every other country. Airline reservation systems, not to mention flight procedures, must be the same everywhere. The quality of goods produced by a small-scale craftsperson, such as a tailor, may be sufficient for local use but may not meet the needs of an exporter who must have uniformity. This quality argument, however, goes both ways. In some cases (e.g., tailored clothing), the handmade item can be of higher quality because it may fit better and include details that the owner specially desires. Products which are handmade are sometimes discriminated against by setting standards unnecessarily high. Appropriate Technology will not be the best solution in every case. In choosing a technology, what is appropriate in a given situation must be determined.

The second concern is whether a nation incorporating Appropriate Technology is doomed to remain small and simple. We have seen that such was not the case in the Far East or Kenya, and U.S. history bears this out — essentially all companies in existence 50 years ago started with a much simpler technology. The historical record appears to confirm that one of the advantages of Appropriate Technology is that it can be an effective way to shift to modern technology. There is no evidence that a country which starts with simple technology cannot move into more complex technology, and there is much evidence that for countries starting with a simple technology the transition to industrialization was easier than it was for those that shifted directly to a complicated case.

One reason some Third World projects have not become more complex may be that aid organizations embrace an approach and continue to use it: For example, "we only fund small projects, using simple machines." The result of such policies is not surprising — projects remain small and simple. This attitude of aid organizations reinforces the suspicion of Third World nations that promoting Appropriate Technology is a way to keep others from becoming industrial rivals.

Actually, in the United States, the opposite problem has occurred: Appropriate Technology projects have become over-complicated. For example, what begins as simple passive solar heating ends up with computer controllers directing the heated air and reflectors tracking the sun. Simple solutions can easily be refined into complicated ones.

Appropriate Technology can be an effective vehicle for preparing people for high technology. One characteristic of Appropriate Technology compared to more sophisticated technologies is that it is more flexible — more choices exist. Appropriate Technology projects need much planning and attention: For example, how can one best use the clay available at a site to make bricks? People at the site must make decisions and take overall responsibility. Planning for Appropriate Technology projects must ensure that the people doing it learn enough both to keep it running and to improve it. When the people operating the technology are in control their expertise and mastery will increase, as will their confidence. Appropriate Technology can lead to further improvements.

# REFERENCES

1. *Appropriate Technology*. IT Publications Limited, 9 King St., London, WCZE 8HW, UK.

   *Published quarterly by a subsidiary of the Intermediate Technology Development Group. Every issue contains examples of Appropriate Technology in the Third World. The basic journal in the field.*

2. Ashe, J. (1986, Summer). Putting your money where your faith is. *New Alchemy Q.* **24**. The New Alchemy Institute, East Falmouth, MA 02536.

   *The article describes the work of ACCION, a nonprofit development agency specializing in assisting local organizations that extend credit and management assistance to "micro" businesses. Their address is 1385 Cambridge Street, Cambridge, MA 02139.*

3. Beckman, R. Presentation on videotape available at cost from the authors at the Division of Engineering, Brown University, Providence, RI 02912.

   *We would be glad to copy it onto a tape sent to us and suggest that persons wanting a copy call us first at (401) 863-2673. We also have copies of the slides themselves and a recording of the narrative which we can copy.*

4. Bordewich, F. M. (1987, March). Development: The lessons of Bhopal. *Atlantic Monthly* **259**(3).

*What are the implications of Bhopal for multinationals in the Third World? Will industrialization be slowed down in India? Are multinationals less careful about safety and the environment in the Third World? The author went to Bhopal to find out and concludes: "Not much," "No," and "No."*

5. Critchfield, R. (1977, September). Small is beautiful, but it's small. *RF Illustrated*. The Rockefeller Foundation, 1133 Avenue of the Americas, New York, NY 10036.

6. Critchfield, R. (1983). *Villages*. Anchor, Garden City, NY.

*Critchfield spent many years visiting villages all over the world, living in them for extended periods, and learning about people's lives. Indonesia seems to have been a favorite place.*

7. Goma, L. H. K., Vice Chancellor of the University of Zambia, used the term in a graduation speech in December 1970 in Lusaka, although it may have been used earlier.

8. Hazeltine, B. (1993, September). Small Scale Technology in Botswana: Utility and Support. *SAMS Working Paper Series*, School of Accounting and Management Studies, University of Botswana, Private Bag 0022, Gaborone, Botswana, No. 93002.

*In the preparation of this report help was received from the following: Agas Groth (Motheo (Pty) Ltd.), David Inger (Rural Industries Promotions), Keith Jefferies (University of Botswana), Khim Kabecha (Bridec), Mmasekgoa Masire-Mwamba (formerly Bridec), R. F. Moamogwe (Pelegano Village Industries), Master Jacob Moelenyane (Rural Industries Innovation Center), and Teedzani Woto (CORDE), all in Botswana.*

9. Kaunda, K. Page 1 of Introduction in *Zambia 1964–1974*, published by the Zambian Information Services, P.O. Box RW20, Lusaka, Zambia.

10. King, K. (1997). *The African Artisan*. Heinemann Educational Books, Ltd., London.

*Describes small-scale enterprises in Kenya, particularly machine making (bicycle parts, maize cutters, and so forth). Discussion about sensible training for such work; very interesting.*

11. Liebenow, J. G. (1981). Malawi: Clean Water for the Rural Poor. *AUFS Reports, No. 40*. Universities Field Staff International, P.O. Box 150, Hanover, NH 03755.

*How villages were mobilized to install their own water supplies, chiefly by collecting ground water from the mountain. Focus is on the politics and organization. Heartening to read a successful project.*

12. MacNeill, J. (1989, September). Strategies for Sustainable Economic Development. *Scientific American* **261**(3), 154–165.

*The section on sustainable development in this chapter is based on this article. Actually, the entire issue is on related subjects and is titled "Managing Planet Earth."*

13. Mavenke, T. (1996). *The Principle and Practice of Campfire*. Campfire Association Publication series, P. O. Box 661, Harare, Zimbabwe.

14. Marsden, K. (1990). *African Entrepreneurs International Finance Corporation Discussion Paper 9*. The World Bank, Geneva.

    *Survey of 36 entrepreneurs from six countries: "Entrepreneurship is alive and well in Africa" and companies do move from the informal sector to the formal sector.*

15. Pacey, A. (1983). *The Culture of Technology*. MIT Press, Cambridge.

    *Good starting place for discussion of nontechnical issues; examples from southern India and from Eskimos.*

16. Program for International Development, Clark University (1989). *An Introduction to Participatory Rural Appraisal for Rural Resources Management*. Clark University, Worcester, MA.

    *A summary of a handbook; emphasis on data collection and on group meetings.*

17. Smillie, I. (1991). *Mastering the Machine: Poverty, Aid and Technology*. Westview, Boulder, CO.

    *A generally optimistic review of aid projects in the Third World, focusing on the effects on the very poor.*

18. Swiderski, R. M. (1995). *Eldoret: An African Poetics of Technology*. Univ. of Arizona Press, Tucson.

    *Gracefully written description and analysis of local technology in Kenya.*

19. The World Bank (1990). *Living with Wildlife: Wildlife Resource Management with Local Participation in Africa* (A. Kiss, Ed.). The World Bank, 1818 H Street N.W., Washington, DC 20433.

    *Projects in Zambia are described on pages 97–102 and 115–122; a wide variety of ways of involving groups neighboring game parks — disrespectfully called "making game guards out of poachers."*

20. Volti, R. (1988). *Society and Technological Change*. St. Martins, New York.

    *Much on the evolution of technologies; pertinent sections on the Bushman of the Kalahari desert and on China. This book has a great deal of valuable material compactly written.*

21. White, L. (1962). *Medieval Technology and Social Change*. Oxford Univ. Press, New York.

    *The example of the effect of the stirrup creating feudal life comes from this book. Other examples include the plow and the mechanical crank; serious and absorbing history.*

# PROBLEMS

1. Use of computers in management has been one of the greatest technological changes in recent years. What are some of the unintended consequences of that

change? What were some of the implementation problems caused by people's attitudes?

2. In the United States why do many young people move to the cities? What has been the effect on the cities and on the smaller towns?

3. Are there equivalents of "colonization of the mind" in the United States, for example, the popularity of foreign automobiles or European clothing?

4. If you were the owner of a small store, for example, a fast-food franchise, how would you develop an attitude in an assistant manager that he or she could solve problems without bringing them to you?

5. What are some possible reasons why aid organizations have not been successful in continuing projects after the aid people leave?

6. Why might Third World leaders not take seriously pleas from the United States to protect the environment?

7. Consider how the norms of behavior in a small community, in which everyone knew each other, would change if a large factory relocated there which would employ more workers than the number of people now in the town.

8. In the United States close family ties usually exist only between parents and children and between siblings. In many parts of the world, these ties extend much further. How is the U.S. pattern consistent with the kinds of jobs wage earners hold?

9. Does a basic American culture exist? Is it in danger of being overwhelmed by foreign music and art?

10. Is the use of high-yield seeds, irrigation, and chemical fertilizer Appropriate Technology? How could its risks to the nation and to individual farmers be reduced?

11. What are the risks in having a community build its own water supply?

12. What are the risks in a development strategy which gives an important place to small producers making umbrellas, wood carvings, lacquer boxes, and so forth for the export trade?

13. What might be the advantages and disadvantages for a Third World company of becoming a partner with a multinational company for local production of a staple crop such as cotton?

14. If you were a business consultant in a Third World country and someone came to you with a proposal to assemble small radios by hand for the local market, how would you respond, i.e., what else would you want to know? No company is currently doing similar things in the country.

15. Some development experts support the "leapfrog" theory that a nation can move directly from low technology to high technology and from no manufacturing to sophisticated factories. Evaluate this theory.

16. Does Appropriate Technology really prepare for other kinds of technology? To be specific, is a self-taught person who has had experience running a bicycle repair shop a better candidate for training as a jet engine mechanic than a person with no mechanical experience and nothing to unlearn?

# APPROPRIATE TECHNOLOGY IN THE UNITED STATES

. . . . . . . . . . . . . . . . . . .

The situation in the United States is different from that of the Third World because much technology is already in place in the United States. Therefore, we ask, can Appropriate Technology work in parallel with other technologies? Because the United States is so large and diverse, it is difficult to decide which are the most pressing needs. Therefore, it must be determined what needs can be most effectively addressed by the Appropriate Technology approach.

In this chapter, we discuss national needs briefly and then we consider three areas in which Appropriate Technology has been useful: a rehabilitation project in North Carolina, inner-city renewal, and an approach to increasing productivity in large organizations. Finally, we consider two concerns about Appropriate Technology in the United States: that it cannot compete with high technology and that it is too far from the mainstream.

## NEEDS IN THE UNITED STATES

Needs include goods and services, jobs, productivity, protection of the environment, and (perhaps) preservation of culture.

Material needs are different in different parts of the United States. Housing is a pressing need in many places, although repairable buildings stand abandoned. In the northern regions fuel for heating is an important need, although the overall supply is currently adequate. Many people are dissatisfied with the quality of services — the streets are in disrepair or dirty, physicians are impersonal, appliances cannot be repaired, and so on. The need for goods and services is paradoxical — there seems to be no fundamental reason why everything needed cannot be supplied but what is actually available seems unnecessarily limited.

Jobs are another need but, again, the situation is not simple. Unemployment coexists with unfilled jobs. One need is for jobs located near unemployed people. Another need is for jobs that are challenging and lead to advancement. Managers are beginning to realize that meaningful jobs are necessary if workers are to do their best. Responsible workers need jobs in which they have some autonomy and in which they can grow.

People who run companies in the United States and those who study such companies are concerned about the relative decline in productivity. It appears (although the data are complex) that it takes more people, compared to our industrial rivals, to make cars or kitchen appliances or other manufactured goods in the United States. It was not always this way. The decline may be caused by the way jobs have been refined and become less satisfying: Jobs on automobile assembly lines, in which each worker is given a very small, routine task to perform very quickly, are frequently cited as overrefined jobs. The productivity decline may also be based on design and manufacturing. An example used often in this regard is Japanese home air conditioners that are improved slightly but significantly every year. U.S. companies, in contrast, kept with an initially superior model too long. Why have Japanese companies been more willing to change or innovate? One answer is that ideas for improvement from Japanese workers and engineers can be implemented easily because responsibility is held by people directly involved with the product, whereas in U.S. companies the responsibility for improvement is held by a senior manager who deals with many products and cannot focus on one. The problem is in the company's organization and planning, not a lack of engineering skill.

Environmental concerns include pollution, loss of open space, and erosion of farmland. A related concern is the heavy use of oil, chiefly as fuel. It seems evident that someday we will run out of oil, that is, the cost of extracting more from the ground will be too high for the value of the oil; people disagree when this day will come. Of course, before this occurs, the price of petroleum will become much higher than it is now and higher prices for gasoline and fuel oil will be a major hardship for most people.

In Chapter 13, we discussed the disruption of culture by technological change as a major issue. In the United States some aspects of culture (e.g., popular music) seem to thrive on new technology. In general, the chance of major, rapid cultural changes due to a new technology seems unlikely in the United States. (For a possible counterexample see Problem 4.) The American cultural heritage actually contains much related to Appropriate Technology: The early settlers were nearly self-sufficient, with simple tools and small farms or workshops. Folk heroes such as Henry Ford or the Wright Brothers, at least in the beginnings of their careers, were small-scale, self-reliant mechanics, and the expression "Yankee ingenuity" refers to the same skills on which Appropriate Technology is based. It is hard to argue that Appropriate Technology is needed to protect U.S. culture, as it may be elsewhere, and in any case Appropriate Technology does seem to be part of that culture. Now that needs have been considered, we examine some examples of successful projects in the United States.

## NORTH CAROLINA DIVISION OF VOCATIONAL REHABILITATION SERVICES

The first of three examples of successful Appropriate Technology projects deals with a way to serve handicapped people. A newspaper article from the Providence *Journal* (October, 1986) (Fig. 14.1) described the problem.

Instead of pouring millions of dollars into new high-tech research centers, the North Carolina Division of Vocational Rehabilitation Services dispatches a small team of circuit-riding engineers to visit disabled people in their homes or work places. The engineers study each person's mobility problems firsthand and then invent a "low-tech" contraption often fabricated from used or discarded materials, to solve the problem.

The solution may be as simple as a brick under a table that permits a client to stay in a chair. The engineers routinely modify wheelchairs, design hand controls for automobiles and lifts for vans, and build ramps for homes and work places. An engineer designed a self-help telephone and special dialing system that allowed a paralyzed salesperson to enter telephone sales. One ingenious solution was a sewing machine that a quadriplegic operates with his forehead.

What the disabled are getting is low-cost, customized equipment they'd likely never receive otherwise. "We've tried to steer our program to the basics — the simple things that enable our clients to live their lives as normally as possible," says George McCoy, coordinator of the Rehabilitation Engineering Program.

**Fig. 14.1. Newspaper article from the Providence *Journal*. (Reproduced with permission of Providence *Journal*.)**

The article states that the program has helped 2500 individuals since it started in 1979. It costs about $500,000 out of the state's $54 million vocational rehabilitation budget. One reason the program is so inexpensive is that state governments pay engineers much less than, for example, defense contractors [7]. Engineers take positions with rehabilitation agencies because the job satisfaction is much higher. The article also comments on the value of face-to-face interaction — partly so the handicapped person works with the engineer to find the best solution. Face-to-face interaction raises the morale of both parties. The engineer sees a person who becomes better off and the handicapped person sees an optimum personal solution evolve.

# INNER-CITY RENEWAL

Successful Appropriate Technology inner-city projects are of several types. We will consider the following: (i) "sweat equity" housing rehabilitation, (ii) assistance in energy conservation, and (iii) urban gardens.

## *Sweat Equity*

Sweat equity results in a person buying a home without a cash down payment. Instead of the down payment the person works on the house, contributing labor (sweat). The houses involved are usually abandoned ones sold by the city for back taxes. These houses have often been gutted to remove anything that can be sold, such as copper pipe or electric wires. The sponsoring agency purchases the house from the city or former owner and then transfers ownership to the new owner. Because of the condition of the house, the price is usually low. The agency arranges financing — a normal mortgage but without a down payment. It gives advice on restoration and makes tools available. It monitors the work for both quality and progress. The homeowner must complete some portion of the restoration before moving in — this effort is really the down payment. In most cases the work is done by the homeowner, basically by hand.

Everyone can benefit from this arrangement. The homeowner gets a home otherwise not affordable and learns skills (carpentry, plumbing, and others). The neighborhood is improved by a renovated house where an abandoned one stood and crumbled, and the city collects taxes. Perhaps the greatest benefit is the pride and responsibility the new owner gains from having rebuilt a house. The benefits are even greater if the new owner had been unemployed — a job and new skills were gained. This approach can achieve better results and cost less than urban renewal through wholesale demolition and replacement.

## Energy Conservation

Assistance in energy conservation, another urban renewal tactic, is much the same as sweat equity. Insulation, tools, and instruction are made available either for free or at cost. Home owners do their own work. Again, they learn carpentry and how to do a project; houses are made warmer and fuel bills lower. The city, therefore, may save on welfare costs. The project is certainly Appropriate Technology: It is labor-intensive, it produces meaningful jobs, it protects the environment by conserving fuel, and the people involved understand it and control it.

Figure 14.2 shows an owner-built greenhouse added to an 80-year-old city home. While the greenhouse was being built and the home insulated, the project was used as a demonstration of energy conservation techniques. Now the building is simply a desirable apartment house near the center of the city.

## Urban Gardens

Urban gardens have been alluded to earlier. The primary benefit is "good food — cheap!" A 20 × 30 ft garden plot can save a family $500 in food costs. A related benefit is that one can grow vegetables not available in the supermarkets. An urban gardener from Southeast Asia got his seeds by mail

**Fig. 14.2. Renovated inner-city home.**

from his home country — actually a risky practice because of the possibility of introducing pests and disease. (The ornamental plant lantana was brought into Africa because someone considered it attractive. The seeds were spread by birds and now are taking over cow pastures.) Two less mixed benefits are the attractiveness of gardens compared to vacant lots or abandoned houses and the bringing together of people in the neighborhood to work on a common project. Urban agriculture flourishes in many parts of the world [9].

The technological problems of starting an urban garden are not difficult. Often a city agency will help carry away debris from the chosen lot. Gardeners have made compost by collecting leaves from parks and streets — garbage collectors are pleased to have someone else take the leaves. Top soil can be purchased. Water must be obtained but nearly all city sites have access to city water. Lead contamination of the soil is a problem as lead poisoning is serious, especially for young children. Lead was formerly used in house paint and in gasoline so sites where buildings or

In the Southside of Providence we have community land where everyone can have a space to plant and garden.

**Fig. 14.3. An urban garden. (Reproduced with permission of Southside Community Land Trust [6].)**

painted structures stood or near heavily traveled roads should be inspected. City health departments or County Agricultural Extension Services can give advice on lead testing and what to do if lead is found.

Starting and running an urban garden requires administration [2]. The land must be obtained. Often the first few abandoned lots are easy to get and inexpensive but after some gardens are established and the area is looking more attractive, the land becomes valuable and desirable to speculators. The group's planning needs to keep this increased land value in mind. Someone has to organize the meetings at which plans are made and plots assigned. Vandalism and pilfering of vegetables do not seem to be a major problem but having someone involved in the gardens living nearby is a good idea. Figure 14.3 shows another view of the garden in Fig. 9.7. Note the burned-out house in the background. Urban gardens have been successful in very unlikely places.

## PRODUCTIVITY

We now consider a third area in which the Appropriate Technology approach seems to be effective in the United States—increasing productivity in large organizations. The aspects of Appropriate Technology that seem most important are (i) simplicity and small scale and (ii) control by the user.

People who have studied organizations note that much seems to be accomplished by small groups working informally, and that complicated procedures and organizations tend to slow things, stifling creativity and responsibility. The expression "Skunkworks," which came from the name of a building at a Lockheed plant near Los Angeles, means a small (approximately 20 person) task force assigned to work on a project with little administrative oversight. Such groups usually are outside of the main structure of the organization. It is surprising that these skunkworks seem to accomplish much more than actively managed groups. Similarly, companies divided into small, nearly autonomous organizations seem to be more efficient than large centralized ones, perhaps because it is easier to get approval for an action if only a few people must be consulted and perhaps because a small organization can substitute face-to-face discussion for written memos. One reason rehabilitation services in the previous example were so effective was because of face-to-face communication. Whatever the reason, people are realizing that keeping an organization small and simply administered leads to better results. Small and simple, of course, is a major principle of Appropriate Technology.

Another important part of Appropriate Technology, with implications toward productivity, is making sure the users understand the technology well enough to adapt it for their needs, such as when the people on the Marshall Islands moved their fluorescent fixtures outside, without guidance from experts. In the Marshall Islands example, however, the islanders could not repair or redesign the batteries, so they really were not in control of the technology. The analogous situation in an organization is whether a group has control over what they are doing: Can the person doing the work improve the process? On some assembly lines, each worker can shut the line down if he or she detects defective parts. In this case the worker has control. Experience shows that people do more when they do have some control. Usually a manager gets best results by only setting goals and guidelines — "We want to build a house in 2 months, using the following material, to these specifications" — and by letting the carpenters, electricians, painters, and so forth decide how to do the actual building. As has been stated, "We want dog owners, not kennel cleaners."

A truly appropriate technology is one that can be mastered by the people using it. Clear evidence exists that the most effective motivation comes from within, not from a supervisor, and people are most motivated to work when they feel challenged but in control. The lesson from Appropriate Technology, that the user must be able to control the technology, holds true in most management situations.

People's preference for simplicity and control is demonstrated by the way services are provided in organizations. There was a time when most people used a computer by time-sharing a main frame. Now, as the reader undoubtedly knows, nearly anyone who has a choice prefers to use a personal computer. A personal computer can do less than a main frame but is more convenient (simpler) to use and the user is not vulnerable to breakdowns about which nothing can be done. Also, a personal computer is not vulnerable to changes in a main frame operating system. The user perceives herself/himself as having control.

To summarize, people who study management have realized in the past few years that organizations structured in ways consistent with Appropriate Technology work better than those structured in other ways. More widespread use of Appropriate Technology ideas could improve the effectiveness of U.S. companies. Many large U.S. companies could be more competitive if they were managed using principles based on Appropriate Technology.

## CONCERNS ABOUT APPROPRIATE TECHNOLOGY IN THE UNITED STATES

Does a place exist for Appropriate Technology in a society full of other technologies? Can it do enough to compete? Is it too far from the mainstream to be worth taking seriously? Can a person survive in the long run using the Appropriate Technology approach? The intention of the examples given in this chapter is to show that situations exists in which Appropriate Technology is more effective than the alternatives. Other examples could have been given, such as improvement of life in rural communities. Appropriate Technology can do enough to compete in certain situations.

To answer the other questions, which ask basically whether Appropriate Technology is a valid alternative, it is instructive to describe the experience of a pair of subsistence farmers in southern Rhode Island. (They discuss these experiences in a videotape available from the authors. For more information see Ref. [1].)

### Subsistence Farming

The names of the farmers are Brian Bishop and Lori Lynn. Their land has been farmed since about 1810. They purchased it a few years ago after renting it for several years. They enjoy living there very much. They grow vegetables — at one time for sale, but now only for their own use. They raise goats, partly for milk and partly as pets. They have chickens for their

SKETCH MAP - AUSTIN FARM ROAD HISTORIC DISTRICT EXETER, R.I.

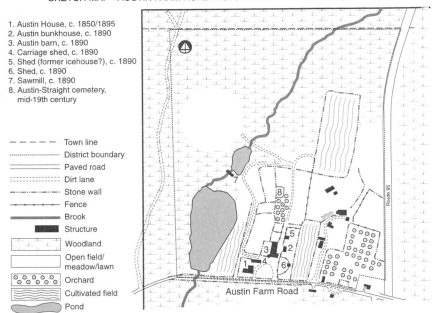

1. Austin House, c. 1850/1895
2. Austin bunkhouse, c. 1890
3. Austin barn, c. 1890
4. Carriage shed, c. 1890
5. Shed (former icehouse?), c. 1890
6. Shed, c. 1890
7. Sawmill, c. 1890
8. Austin-Straight cemetery,
   mid-19th century

— — — — —  Town line
·······················  District boundary
══════════  Paved road
··············  Dirt lane
—··—··—··—  Stone wall
—•—•—•—  Fence
▬▬▬▬▬  Brook
▬▬▬▬▬  Structure
☐  Woodland
☐  Open field/
   meadow/lawn
☐  Orchard
☐  Cultivated field
☐  Pond

**Fig. 14.4. Brian and Lori's farm. (Reproduced with permission of Brian Bishop.)**

own use. They board horses and Lori gives riding lessons (Fig. 14.4). To earn cash for the mortgage payments Brian repairs houses approximately 35 hours a week for a real estate management company in a nearby city.

They grow vegetables organically and have several deep freezers to keep vegetables through the winter. For 2 years, they grew vegetables on a 2-acre field and sold them from a pony cart in the city. Sales were good and people liked the pony. The time spent minding the cart took them from more profitable activities so they gave up the cart and the large vegetable garden. Brian says that even now they grow more vegetables than they can use and he laments the lack of a convenient way to transport vegetables to a soup kitchen. The one large crop they do raise is hay for their own use. For this, they use chemical fertilizer. The price charged for hay grown in Vermont is about the same as their costs; evidently the costs of Vermont hay are much lower than those in other states, so it does not pay to raise hay for sale.

The horses are the big cash earner. Brian and Lori board about a dozen horses and own eight others. Lori enjoys being around horses and giving lessons. They also have a dozen or so goats and use the milk themselves. They sell some milk but more production would require milking machinery and refrigeration and, if the operation grew large enough, state regulation. The goats are really pets and are tame and affectionate. A few goats have been sold but the demand is generally small.

There is an old hydropower installation on the farm and Brian would like to get this running again. He also is considering a summer farm camp for children, perhaps located in the mill house. At times various volunteer workers have stayed with them to learn about farm life. Because the farm is one of the oldest continually functioning farms in the state, he has applied to have it designated as a National Historic Site. Such designation would have tax advantages. If a time comes when he does not need the cash from city work he would like to get back into growing vegetables for sale and into goat milk production.

Brian thinks about how self-sufficient he wants to be or can be. He needs electricity not only for the freezer but also for the water pump and lights and appliances. A major concern is what he calls his support network. He shares haying equipment with another farmer. If that farmer goes out of business, as many have done, Brian will be in trouble. If he needs repairs on most of his equipment, he must go to a machinist or a salvage yard because none of his equipment has been produced in a long time. What worries Brian is that the salvage yard is run by someone over age 70 and the machinist is older. Another neighbor, also about 70 years old, runs a saw mill. Brian could survive the loss of the saw mill but the other two people are important to his operation. Brian says he got worthwhile advice from several older farmers when he started farming. The state agricultural extension officer does not deal with his kind of farm. Even a veterinarian knowledgeable about goats was hard to find. Brian stresses how much knowledge is needed to run a farm—from plants to animals, and government policy. One reason Brian enjoys the life so much is the intellectual challenge.

What good are Brian and Lori accomplishing? One reason they were able to farm was that the town refused to let the previous owner cut the property up into house lots. In fact, houses are currently being built across the road from them. Therefore, the farm is preserving open space. It is also providing services—recreation, open space, and some food—and when they are more established, they may be able to provide more services. They are learning how to be successful subsistence farmers so it will be possible for others to choose to do the same. They are maintaining part of the national culture, the culture represented by both the self-reliant farmers and the ingenious artisans that were important in American history.

## JUSTIFICATION FOR APPROPRIATE TECHNOLOGY IN THE UNITED STATES

One lesson we can learn from Brian and Lori's situation is that just because they are doing something different they are being useful. The concern that Appropriate Technology is too far from the mainstream to be useful really should be a concern that diversity be maintained in the United States, that choices remain, and that methods that were successful not disappear. Some things can be done best through Appropriate Technology in the United States today. The country as a whole would be poorer if Appropriate Technology-like activities disappeared.

## REFERENCES

1. Bishop, B. and Lori Lynn's video is available at cost from the authors at the Division of Engineering, Brown University, Providence, RI 02912.

   *We would be glad to copy it onto a tape sent to us, and suggest persons wanting a copy call us first at 401-863-2673.*

2. Boston Urban Gardens (1982). *Handbook of Community Gardening*. Charles Scribner's Sons, New York.

   *Straightforward advice, mostly on organizational problems — getting the group together, getting help, building community organizations.*

3. Marx, L. (1964). *The Machine in the Garden, Technology and the Pastoral Ideal in America*. Oxford University Press, New York.

   *Probably the most often referred to book dealing with the role of technology in American culture. It is probably best to review* Huckleberry Finn *by Mark Twain before reading Marx. Twain approves of technology, Marx doesn't.*

4. McRobie, G. (1981). *Small is Possible*. Harper and Row, New York.

   *Part Two, pages 75–179, titled "Technology Choice in Rich Countries," reviews groups and projects in England, the United States, and Canada. Addresses are given. Names and the emphasis have changed since 1981 somewhat but the overall picture is much the same.*

5. Peters, T. and Waterman, R. H. Jr. (1984). *Search for Excellence*. Warner Books, NY.

   *This was a bestseller for several years. It is based on observations of successful companies and aims at recognizing the factors leading to success. A strong plea is made for simple structures and freeing people in the organization to do their best.*

6. Schimberg, D., *Southside Green*. Newsletter of the Southside Community Land Trust, 305 Dudley St., Providence, RI 02907.

*Most of the urban garden pictures in this book are from this project. The Southside Community Land Trust has 250 garden plots, raises chickens, gives classes for children, and does much more. The telephone number is (401) 273-9419.*

7. Smart, E. D., Rehabilitation Engineer, Vocational Rehabilitation Services, 743 Fifth Street, SW, Hickory, NC 28601.

   *Has articles and information about the program.*

8. SWAP Housing Program, David Karoff, 439 Pine St., Providence, RI 02907. Phone: (401) 272-0526.

   *The SWAP program purchases abandoned houses and helps new owners find funding for renovations. They also give advice on how to do renovation. Analogous groups exist in other cities.*

9. United Nations Development Programme (UNDP) (1996). *Urban Agriculture, Food, Jobs, and Sustainable Cities.* UNDP, One United Nations Plaza, New York, NY, 10017.

   *Extensive review of activities across the globe; policy oriented; many contact addresses.*

# PROBLEMS

1. Some people blame changes in technology for many of the United States' problems, for example, unemployment is blamed on factory mechanization. Choose a national problem and show how technology caused or at least exacerbated it.

2. This is a companion to No. 1. Choose a national problem (e.g., literacy) and show how technology could help solve it. Consider several different technological approaches to the solution.

3. Choose an American novel, such as one by Mark Twain, and consider the role played by technology: Was it important, beneficial, and to be nurtured or feared?

4. People have argued that the sexual revolution, more freedom in sexual behavior compared to previous generations, was created by technological change — contraception, antibiotics, automobiles, and motels. In what ways would you agree?

5. Consider another social service besides rehabilitation which might be performed better if a less elaborate technological approach was used (maternity care is an often-used example).

6. One concern about self-help urban renewal projects is gentrification: middle-class people move into run-down neighborhoods to live and they improve the housing and the neighborhood and thus prices increase. Who benefits from gentrification? Who suffers? How can the hardships be reduced?

7. Analyze the claim that a person who has rehabilitated her/his own home will therefore also contribute to the community in other ways.

8. A problem with Appropriate Technology in the Third World has been that research groups become isolated and cannot make use of information developed elsewhere.

What is an analogous problem for organizations split into autonomous task forces? What can be done about the problem?

9. Discuss other ways Appropriate Technology can contribute to life in the United States, for example, in the areas of transportation, health care, or urban crime.

10. Brian and Lori work on their farm because they enjoy living in the country and like animals. They support themselves but do not add a great deal of cash to the state's economy. What position would you take if an electronics company wanted to take over their land, which is inexpensive but accessible, to build a factory? Consider the issue from the point of view of another resident of their immediate area, from the point of view of a city or suburban resident, and from the point of view of a state economic planner.

11. How could an organization interested in preserving parts of American heritage most effectively support Brian and Lori? What are the risks in simply making cash grants?

12. Decide for yourself where to draw the line for self-sufficiency on a farm such as Brian and Lori's. Would you buy electricity? Would you eat only meat you raised yourself? What about clothing? What about chemical fertilizer? What about shots for the animals?

*Chapter*

# 15

· · · · · · · · · ·

# CULTURE AND WOMEN

· · · · · · · · · · · · · · · · · · · ·

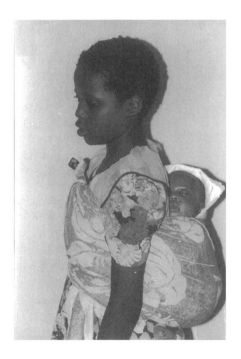

## THE RELATIONSHIPS BETWEEN CULTURE
## AND TECHNOLOGY

Two relationships between culture and technology are important. One is how a particular technology can augment or diminish a group's culture. The other is how cultural attitudes determine whether a new technology will be accepted by a group. Both of these relationships are important if we are proposing a change in technology. It is irresponsible to propose changes without analyzing the effects of those changes, both materially and culturally. It is wasteful to propose a change that will be rejected.

By the word "culture" we mean a group's customs and standards of taste. Culture includes the norms of how people live with each other—their attitudes about generosity, honesty, and courtesy. It also includes their arts—literature, dance, painting, carving, and clothing. A major point of this chapter is that, although intangible, these cultural attributes are important.

The world would be a much poorer place without a variety of cultures, in particular if Third World cultures were homogenized into Western culture. Examples of how foreign cultures can enrich our lives are easy to find. Visitors to many parts of the Third World are impressed by the courtesy and warmth with which they are greeted. Sometimes this warmth can be embarrassing: In the Middle East one must be careful not to admire something too enthusiastically, lest it be presented as a gift [14]. Another point of this chapter is that insensitive technological planning can do harm to valuable local customs and attitudes.

A major question when considering culture and technology together is whether one can gain the good aspects of one culture without also getting the bad. To be specific, one person's list of good aspects of Western culture might include a sense of individual worth, of boundless opportunities, and of responsibility for one's actions. The same person's list of bad aspects might include neglect of the aged and disregard of the environment. Is it possible to have a society with these good aspects but without the bad? Can use of Appropriate Technology make it more likely that change will have a net positive cultural effect?

## Technological Changes Can Produce Cultural Changes

The arts are a good example of how technology has influenced culture. Art forms have been greatly changed by new devices or new materials. Electronics made movies, TV, and radio possible. Photography, printing, and even the personal computer have provided many more choices in decorating the walls of a room. Steel frames and reinforced concrete made new kinds of buildings feasible and thus changed architecture significantly.

Technology can have a negative influence on the arts also [1]. People used to gather around a piano and make their own music. Now they turn on a stereo. Electronics have diminished the level of musical skills and made the group less cohesive. Of course, the stereo is easier to play—benefiting the musically inept — and the choice of music is wider because many more selections are recorded than most amateur musicians can play. On the other hand, availability of recorded music can cause a loss of cultural diversity. If the only recordings commercially available in a Third World country are of foreign music, because the country cannot support a recording industry, then local music may die out. The situation, however, is complicated. Availability of inexpensive, quality recording equipment may, in fact, make local music more visible and give it stature. In the United States it is questioned whether the increased number of radio stations and sophisticated recording equipment has helped or hurt regional musicians. We can say that in both the United States and the Third World the potential influence of electronic technology on local music is significant and can be harmful.

The coming of a new technology can affect entire ways of living. One of the reasons Europeans came to Southern Africa was to create a mining industry. Mining operations had existed before the Europeans arrived but the mines were small, needing only people living in the nearby village to work them. The early European prospectors actually based much of their searches on what local people had learned about mineral deposits — they asked old people in villages where ore had been obtained in the past. The Europeans, of course, dug large mines using mechanized technology. To operate these mines many more people were needed than a village could supply so workers had to be brought in. In order to induce people to be miners, as well as for other reasons, colonial governments imposed a hut tax on African communities. The tax had to be paid in cash, which was scarce because it had been seldom used in the villages. Young men thus were forced to leave their villages in search of paid employment and many ended up in the mines. Three or 4 months of mining work was usually sufficient to cover the hut tax but some men elected to remain in the mining towns, usually in shantytowns on the outskirts. Perhaps the mining towns seemed to offer more opportunities for advancement than the villages; at the least they were a way to be independent. The colonial government did not want to provide housing, education, or other services for unemployed persons because their intention had been that once the tax was paid people would go back to their villages. Living conditions, therefore, were poor for most African men in these mining towns because of the absence of government services and few brought their families to try to make permanent homes. The presence of many unemployed people living in makeshift accommodations made the mining towns less attractive. The villages were also worse off because of the loss of often the strongest and most ambitious men. The result was that cultural life, both in the villages and in the mining towns, was diminished: The villages had too few men and the towns had too many. The new mining technology had deeply affected the culture [6].

## *Cultural Factors Can Influence Technological Change*

A cultural factor that was not always taken into account when economic planning was done in Africa is the responsibility that many people feel to their extended family. In some of these cultures it is customary, when one person has found a job, for unemployed family members to move in and share the good fortune. Such a custom has obvious advantages for family unity but equally obvious disadvantages for encouraging individual accomplishment. Why should a person learn a new skill or take more responsibility in a company if most of the added benefits will go to relatives? A program in Malawi which trains small-business people to repair radios or make roof tiles from cement recommends that new enterprises be established in places some distance from where the entrepreneur's extended family lives. If cultural factors are not considered, these new businesses will have difficulty.

Some technological changes have been thwarted by factors we think of as being "merely aesthetic." Improved food production is an illustration. Soy beans, in a sense a new technology, are more nutritious and do more for the soil than other crops but they take longer to cook and the taste is different — they are less culturally desirable. Because of the taste and the preparation time they have not been widely accepted despite their good features. Yellow maize (corn) flour produced in abundance in the United States is only eaten as a last resort in Central Africa, where the white variety is preferred. Cultural attitudes cannot be ignored even when the new technology seems to offer more benefit for less cost.

Firewood conservation is an example of how women influenced technological change. The World Bank, to protect forests, has put a major priority on decreasing the consumption of firewood. Part of this effort has been the promotion of a new technology — a more efficient wood stove. These stoves have not been accepted everywhere. Observations from both India and Zimbabwe indicate why [15]. Because these stoves are more efficient they cook slower and use less wood — saving time spent gathering firewood. The women involved, though, consider cooking time more significant than fuel consumption because wood gathering is an opportunity to socialize, so the time spent is not considered wasted. Efficient wood stoves did not meet a cultural need and were not accepted.

The interaction between culture and technology is a broad topic. We will primarily focus on women and Africa. We will examine how technological change affected women, comparing their situation to that of men. Then we will examine how Appropriate Technology can help make a change in technology beneficial to women. While the discussion is about Africa, the issues are also important in the United States [19].

# WHY WOMEN?

Women were chosen for this discussion because although they compose about 50% of the population, they produce, in Africa, between 60 and 90% of the food, process over 90% of it, and carry nearly all the water and fuel (Fig. 15.1). Until recently, however, not much has been written about how technological change affects their roles and lives. This lack of writing probably reflects a lack of consideration of the issue — perhaps a belief that if the general level of technology in a village is raised then life will improve correspondingly for women. Such uniform improvement has not always occurred and cultural factors have been one reason.

Well-meaning but imposed changes can make life worse for women. In the wood stove and soy bean examples, women could refuse to accept a technological change. In other cases, the change occurs no matter what women want. The purpose of the change may be to help the community but unintended consequences may actually make life worse. A new housing project which had homes separated some distance from each other meant that mothers could not conveniently watch each other's children when one of them was gone. A design intended to help, by giving privacy, in fact made life more difficult because the design did not match the sharing culture.

**Fig. 15.1. Women in an African village.**

Special attention should be given to how a technological change affects women because women nearly always are the ones taking care of infants and children. Cash raised by women usually goes into meeting household needs and much of this cash is used on infants and children. A study in four African countries [7] showed that time freed up by the introduction of new technology was spent on improving life for family members. Changes that give an increased burden to women, such as those that take men out of the household or make it harder for women to earn cash, can mean that food and health care for children will be diminished. Children's education is also important. Obviously, long-term improvement in people's lives requires education. If women are overburdened, they cannot tend to their children's education. Thus, danger is that technological change will make nurturing the young more necessary but that nurturing will be less attended to by women.

## How Technological Change Affected Women

In general, the pattern of change has been that new technologies brought better jobs for men but not for women. We consider an African village. Originally, much of the work was shared, although some, such as hunting, was done mostly by men and some, such as making household items, mostly by women. When new technologies arrived (e.g., mining and manufacturing), either the women's jobs disappeared and were replaced by inferior ones or women became entirely responsible for the household as men left [18]. Women's jobs disappeared if the family made a transition to a cash economy, depending on wages earned in the new industries to purchase household items. In this case the family did not need the things women had been producing but did need cash. Therefore, many women had to take menial jobs, often as domestic servants and nearly always outside the home, to earn money. If, on the other hand, the men left the village to take jobs in mining or manufacturing towns, women had to perform all the family and community tasks. In either case, women were worse off.

Development projects, for the most part, have not helped women as much as they have helped men. Usually when new farming methods were brought in, the improved tools and training, as well as the new seeds, were aimed at large operations rather than domestic gardens. Little attention has been paid by most aid agencies to the particular technologies that require the most time for Third World women (e.g., home-scale food preparation or storage). Aid projects appear to have been planned in the belief that women and men would share equally when agriculture or manufacturing was improved, but this belief has often not been true. In the next section we consider why women and men were treated differently when technology was transferred or changed and why development projects often focused on the technologies most useful to men.

## Why Technological Change Affects Women Differently

Two reasons exist for why men benefited more than women when new technologies were brought to Africa: traditional heritage, based on life before the Europeans arrived, and colonial heritage, based on the attitudes and actions of the European settlers.

To understand the traditional heritage we should consider in more detail the traditional village. An early explorer [6] wrote that village men spent their time hunting, fishing, and warring. Women, as part of family responsibility, gardened, chiefly for domestic consumption. Population pressure was slight, partly because infant mortality was high. Small numbers of people meant that much food could be gathered from the wild so nobody had to work more than 3 or 4 hours a day. Although job responsibilities were divided by gender, much job sharing existed, especially since there was usually a surplus of men. Wood carving and basket making became male tasks, perhaps because women were otherwise occupied.

The men took different roles with the establishment of a European administration which stopped wars and curtailed hunting. These new roles taken on by men were often the judicial and ritual ones. In present-day Africa, most herbalists (traditional doctors) are men (Fig. 15.2). Men may have become herbalists because their traditional responsibilities had disappeared; this may also be the reason why they assumed the judicial and ritual positions, although these roles may simply have evolved from tribal chieftainships.

For whatever reason men became judges, the judicial role became important in areas administered by "indirect rule," as the British used, in which local self-government (reporting to colonial officials) was instituted. This self-government included a separate judiciary based on the British model. The common practice of having nominal land ownership in the hands of men may be a result of having male judges and community spokepersons — land ownership did not appear to exist in precolonial Africa. The result was that men's role expanded with the colonial era.

Colonial policies had other results. Colonial administrators tended to deal with men exclusively, perhaps because of their own European traditions. They did not encourage women to participate in local government nor to try to become landowners. When early development projects were

**Fig. 15.2. Present-day herbalist.**

instituted, men were the ones trained or involved in planning. When factories and plantations were opened, men were given the responsible positions.

As a result of colonial policies a general pattern emerged that men performed the tasks requiring interaction with outsiders and women performed the local tasks. City markets in Malawi reflect this structure today. Fish and produce that come from some distance away are sold by men. Locally grown produce and homemade utensils, such as clay pots, are sold by women.

Technology transfer sometimes favored men in another way. Machinery sent to the Third World was often the wrong size for women or otherwise unsuitable. A recent example is a water pump brought to the Sudan which was unusable by women because it required the operator to straddle it, which no self-respecting Sudanese women would do. Furthermore, many development projects were primarily aimed at increasing production for items in the cash economy (items usually made by men). Colonial administrators and development officers acted, in a sense, as gatekeepers, sending new technology to men and ignoring women.

Men's jobs were upgraded and women's jobs downgraded as a country industrialized. Some men were encouraged, because of industrialization, to become machinists or building technicians, which often meant a sizable cash investment for tools. Cash was diverted from domestic needs to the new jobs. On the other hand, new factories could inexpensively make items women had been making at home: an often cited example is a plastic sandal factory producing sandals at a fraction of the cost of the leather ones women had been making at home [2]. Plastic sandals use imported raw materials and are not repairable. Therefore, in the long run they are less desirable but the initial cost is lower and the sandals appear modern. The plastic sandal factory takes jobs from women and, because the women's income often pays for family expenses, has made both women and children worse off. As a result of industrialization, therefore, men in the household may have high cash needs — to compete in the cash economy — and women may have lower cash sources because their products are no longer competitive. Furthermore, work has tended to move out of the family — less of it is shared. The former approximate equality between men and women in the family has disappeared. This disappearance has especially affected women in societies in which women cannot own property because without collateral they have difficulty raising funds to improve their position (e.g., to open their own sandal factory).

This change in the relative positions of men and women in a family can also be seen in agriculture. High-yielding varieties of seed — the Green Revolution type — can be a mixed blessing to women, especially if only men are given the new seeds. When a woman has poorer seeds, she earns less from her garden. On the other hand, the man's garden requires more work, high-yielding seeds need better treatment and care, such as irrigation and fertilization. Women may be called to help with this care. A field of high-yield grain tends to ripen over a short time, which is helpful for mechanized harvesting but a burden on a hand operation because of the urgency in getting the crop in. Women, again, may be pressed into service during this crucial time. High-yielding seeds thus have made the woman's garden less valuable and have increased the burden on her. Equality has been replaced by dependence. What appeared to be a beneficial policy (new seeds) may not improve family life because of cultural factors.

It should be pointed out that since the former colonies in Africa became independent nations, they have made strong efforts to employ women in policy-making positions. Aid organizations are also recognizing the need to address women's issues explicitly, as described in the next section.

## What Can Be Done to Make Technological Change Beneficial to Women

We focus here on activities with tangible outputs rather than ones that primarily aim to change cultural attitudes, although both are important in meeting women's needs. Redesigning existing equipment and ensuring that new equipment is suitable for women is one important way of adapting technology. Tractors have been sent to the Third World with steps too high for women to climb onto comfortably or pumps "that couldn't be used by anyone with breasts" [8]. A project that would appear to be very helpful to women is the design of an efficient wood stove — a project undertaken by the World Bank and others. Often neglected in these designs are the needs of the user, such as reduction of smoke or providing warmth to the room as well as fast cooking. Keeping designers in touch with actual user needs is not, of course, a problem unique to wood stoves. Perhaps it is especially severe in this situation because the designers are mostly male and on a different continent than the user.

Another important step in making new technology useful for women is providing women access to education. The belief that educating only males results in well cared-for families has turned out to be false. Education for women facilitates cooperation between women and men. Also, education for women results in improved care for young children [3]. Of course, education makes both women and men more productive and thus improves the community.

In addition to education, it is beneficial to make it possible for women to borrow money, giving them the opportunity to start new businesses, which bring new technologies into a community. The products of businesses started by women tend to match women's needs well. The experience has been that the cash earned goes directly toward meeting family expenses.

Special programs to develop and bring technologies to women in the Third World can be effective. The United Nations has recognized this need and set up a program called UNIFEM [7]. Conferences for planners have been conducted, that concluded with one in Nairobi in 1985. Workshops for village-level trainers have been held, and guides have been published (the cover of one is shown in Fig. 15.3) [20].

## How Appropriate Technology Can Help

The characteristics of Appropriate Technology — low cash requirements, use of locally available resources, repairable by user, and controlled by user — are well matched to the situation of women in the Third World. We have seen that often Third World women do not have cash and thus cannot buy imported food and clothing. They have usually had less schooling than men so the equipment they use must be maintainable with little training. Because they often have to care for children, their jobs must be flexible. Several specific Appropriate Technology projects that have met the needs of women will be considered in this section.

Women in the Third World need to free up time, partly so infants and small children can be tended. A project in a Nigerian community did this

Prepared in conjunction
with a
UNICEF
FISH SMOKING
EXTENSION PROJECT
of the
NATIONAL COUNCIL
ON WOMEN
AND DEVELOPMENT
assisted by the
FOOD RESEARCH
INSTITUTE
carried out in coastal Ghana
(Greater Accra, Ada and Keta)
from June 1983 through January 1984

Based on improved fish smoking
technology developed by FAO, the
Food Research Institute and the
Women of Chorkor, Ghana 1969–71,
and widely utilized in the area.

*Sardinella*

**Fig. 15.3. Cover picture of a manual for an improved fish smoking oven. (Reproduced with permission of UNICEF.)**

and earned money. When a child is being weaned special foods are often needed because adult foods may be too spicy or too solid to be digested. These weaning foods should be higher in protein and calories than adult foods because the portions are smaller. In this community, many mothers worked and kept their children nearby, feeding them during the noon break. A project trained food vendors (mostly women) to prepare, at suitable times, bean and peanut meals for children of working mothers. The project also provided loan money to the food vendors to start the business. These weaning foods turned out to be a profitable side line. The vendors gained a new business and the working mothers nutritious baby food. An alternative approach considered was to educate mothers to make special weaning foods from local cereals, legumes, and oilseeds. This approach was dangerous for working mothers, however, because diarrhea-causing pathogens rise to dangerous levels in food prepared early in the morning and kept in the heat during the day. Also, preparation would have had to be done in parallel with the mother's job. Training vendors was a more effective strategy.

A project in Papua New Guinea [5] met cash needs for women with deep-fried banana chips (banana chips taste like potato chips). A village group produces the chips and sells them commercially. The project made use of much Appropriate Technology hardware—a Lorena stove, a ferro-concrete tank to collect rainwater for washing, and a bag sealer made of a candle and a hacksaw blade. The kitchen was built by village people. The project employs five skilled women cooks, a woman manager, and several young boys and girls. The earnings from the project go to the village women's cooperative. The cooperative could not have paid for expensive machinery: Without Appropriate Technology this project could not have been done.

Appropriate Technology provides other opportunities for women to earn money. The tailor shop in Fig. 15.4 is owned by a woman who essentially coordinates the work of several tailors making custom dresses at home. Handicraft cooperatives exist in many parts of Africa which purchase homemade curios, jewelry, or embroidery from women and sell them either in cities or to exporters. Handicrafts can be made at home part time. The risk, of course, is that making things for sale becomes another burden on already overworked women, but it is possible for handicrafts to be a cash source that does not interfere with other tasks.

The possibility that Appropriate Technology will in fact result in more work for women should not be disregarded. A report [9] noted, "[Does] the move to simpler technologies in the industrialized countries mean back to the kitchen for women?" The answer is not simple. Historian Ruth Cowan Schwartz [10] points out that despite major technological changes in the household—vacuum cleaners, dishwashers, clothes dryers, and so forth—the number of hours spent by present-day U.S. housewives doing housework is about the same as that spent by their colonial counterparts.

**Fig. 15.4. Tailor shop in Nigeria.**

Perhaps expectations have been increased by the new technology. Mechanization has not reduced the number of hours spent on housework. It is hoped that less mechanization will not result in more hours.

The previous examples (weaning foods and women-owned-small businesses) showed how the Appropriate Technology approach could improve the lives of women. Now we will look at examples showing how Appropriate Technology has supported existing cultural patterns during changes.

## APPROPRIATE TECHNOLOGY AND CULTURE

The reason Appropriate Technology relates well to cultures is that it adapts to local needs and it is controlled by those using it. A classic book, *Behind the Mud Walls 1930–1960* [20], written by two American missionaries who lived in India in 1930 and one of whom returned in 1963, describes how village life changed during this time span. The major change ("We have more to eat and more to sell") was the result of a technological change — better seeds, fertilizer and compost, and new hand tools. The national government developed new techniques and sent extension workers to demonstrate them and to assist and advise villagers who wanted to try them. Villagers "with enough land and enough food in the storeroom to risk a field" experimented with the new seeds and cultivation techniques. The rest of the villagers saw what worked and copied it, rejecting or modifying what did not work. Much worked and for the first time significant amounts of money came into the village. With more money, more wells were dug and water was generally available. Trained physicians settled in the village. Young people, especially, were optimistic about the future. However, traditional patterns of life did not markedly change. People still supported the family, were willing to join with other village members in times of crisis, and gained satisfaction from religious festivals. The new farming technology, which was directly related to the former technology but more productive, had made life better without exacting the price of major disruption.

One group did not prosper with better farming. These were the craftspeople — carpenters, textile workers, and the people who pressed oil out of mustard seed. These people found they were losing business to cheaper or more versatile machines in the nearest city. The technology question is whether improved, inexpensive tools can make these people competitive. Would Appropriate Technology have made it possible for these people to remain in the village, continuing with their work? Another question, of course, is "Why care?": Is a village significantly worse off without craftspeople?

Changes in village life over the 30 years the missionaries observed seemed to have occurred easily. The authors were impressed with how much had been accomplished. How was a major improvement made in people's lives without serious outside pressure on them? It is hard to tell but perhaps the Appropriate Technology approach—hand tools, local control, and adaptation to local conditions—was one reason. Another reason may have been that the government agents working toward change were all Indian—people who probably understood their audience and realized how to be most convincing.

What can we learn from this Indian village about the relationship between Appropriate Technology and culture? Promotion of labor-intensive farming techniques meant that agricultural jobs were not lost. The new tools and methods were similar enough to what had been used and done before that major changes were not forced on the society. Because the new techniques were fairly simple it was possible for villagers to select which improvements they would implement; thus, the risk was reduced that the new methods would result in a disastrous crop failure. Because the advisors were close to farmers in the village they were able to recognize how the new farming techniques matched the attitudes and needs of the farmers as well as matched local soils and rainfall. The new farming techniques mostly used tools and supplies that could be repaired or found within the village so the village maintained its identity. The success of the farming improvements contrasts with the harm done by wood-working machines, textile equipment, and oil presses in the city, which made village artisans redundant. Appropriate Technology made life better for village people and was gentle on the culture.

## APPROPRIATE TECHNOLOGY AND MAJOR CULTURAL CHANGE

In some situations major cultural change may be to the benefit of the people involved and they may desire it. We have taken the view that any induced change in cultural norms is suspect. One should certainly show respect for different cultures and this respect was often lacking in the past, but an attitude which discourages any form of change can be callous. It might prevent, for example, drilling wells in a community because women carrying 10-gallon water buckets on their heads are seen as picturesque. What an outsider may think worth preserving may not be at all what those involved want to keep.

An article [12] about the Rendille tribe in northern Kenya shows the difference between the attitude of the people and that of an American anthropologist. The Rendille up until 1975 were animal herders — camels and cattle mostly. They would settle in an area until the grazing ran out and then move on. During one of the periodic famines a relief organization began distributing food, corn meal, and powdered milk at a town called Korr. The Rendille settled nearby to survive the famine. Their animals quickly overgrazed the surrounding area. Many animals died and the rest were taken back into original grazing lands by young men and, as the famine came to an end, the herds slowly regained their strength. Most of the Rendille, however, stayed in Korr, without their animals, and depended on the food given by the relief agency. The agency had built wells, a primary school, and a dispensary. To the anthropologist, life at Korr seemed dismal and lacking in purpose compared with life at the nomadic settlements, but a woman at Korr told him "here at Korr there are shops close by, school for my children, a dispensary if you get sick. And when I need water it is right down the hill at the pump. I hope I never go back." The point of the example is that the advantages from the improved technology, water, food, medical care, and schooling, seemed more important to the Rendille than the cultural loss, at least in 1980.

Actually, the situation was more complex than just described. Children suffered more from malnutrition in Korr because the diet was more limited than that in the grazing areas. Because the town was overcrowded infectious diseases were also more common. What had begun as straight-forward famine relief efforts caused major changes in people's lives.

What if a major transition in a community's culture is desired? Is Appropriate Technology limited to situations in which only a small change is desired, as in the Indian village previously described? Major social changes have, in fact, come from small-scale, decentralized, under-funded efforts: The environmental movement is such an effort. Not only does the national concern for the environment have a heritage consistent with the Appropriate Technology approach but also national leaders are depending on a similar approach to protect the environment. Individual families are being called upon to do relatively simple things on their own, such as insulation or sorting garbage. One hope is that educating people about things to be done will change their attitudes about energy waste or trash disposal. The Appropriate Technology approach is being called on to assist in a significant change in national culture.

# SUMMARY

The basic concern in this chapter, as elsewhere, is to find new technologies that improve the lives of people involved. The process of making mean-ingful technological change is not only a technological one. If cultural factors are not taken into consideration when introducing a new technology it is likely that the aims of the new technologies will not be met, perhaps because of unexpected consequences or perhaps because of resistance by those involved. We have discussed examples in which the aims were not met because women, who are prime agents in social change, were excluded. We have also discussed examples in which the aims were not met because the technological change was so severe that the culture could not adapt, such as when centralized industries were introduced in Africa or nomadic tribes were forced to settle in one place. Appropriate Technology tends to put participants in control. It can be adapted to local conditions. It does not require major changes in people's lives. It is a promising way to improve living conditions without cultural damage. It is also a promising way to make major cultural changes if these are desired.

## REFERENCES

1. Abu-Lughod, L. (1987). Bedouin blues. *Nat. History* **96**(7).

   *The author lived among Egyptian Bedouin recently. Her primary focus is on love poetry written by individuals, who have no other socially acceptable way of communicating their feelings. Poetry readings on cassettes affected this poetry significantly. The article also has much to say about what happened when these nomadic people settled.*

   *Nearly every recent issue of* Appropriate Technology *contains articles dealing with women. The following four articles were especially helpful in preparing this chapter.*

2. Carr, M. (1978, June). Appropriate Technology for women. *Appropriate Technol.* **5**(1).

   *The lead-off article in an issue devoted to women; chiefly about how new technologies affected women in Africa. This issue of* Appropriate Technology *is a good starting point for further study of women and Appropriate Technology.*

3. Appleton, H. (1993, September). Gender, technology and innovation. *Appropriate Technol.* **20**(2).

   *The lead-off article in a second issue devoted to women. Focuses on how external factors influenced the development of technology and how the external factors were quite different for women and men.*

4. Fleming, S., and Margoram, T. (1986, June). A South Pacific workshop. *Appropriate Technol.* **13**(1), 12–13.

   *Discusses the topics covered in a 6-week program for development trainers and extension workers, "who will transfer useful skills to village women": small business management, nutrition food processing, energy, water supply, and sanitation. Article is a bit skimpy but idea is good.*

5. Rashinah, N. (1984, June). The Situm banana chip enterprise. *Appropriate Technol.* **11**(1), 28–30.

*The enterprise was deep frying green banana chips and packaging them for sale at a supermarket. It took place in Papua New Guinea and was successful; thorough article.*

6. Cairns, H. A. C. (1965). *Prelude to Imperialism, British Reactions to Central African Society, 1840–1890.* Routledge Kegan Paul, London.

*Includes reports, often contradictory, showing how local people looked to the explorers, missionaries, and settlers who came first.*

7. Carr, M., and Sandhu, R. (1987, September). Women, technology, & rural productivity. In *UNIFEM Occasional Paper* (No. 6), UNIFEM, 304 East 45 Street, New York, NY 10017.

*Brings together much research on how women in the Third World spend their time — mostly working very hard, certainly harder than men.*

8. Margaret Catley-Carson quoted in IPS press release, New York, December 24, 1989.

*Margaret Catley-Carson was president of the Canadian International Development Agency.*

9. Women's Resource Center (1979, April). *Conference Proceedings — Women and Technology: Deciding What's Appropriate.* Women's Resource Center, University of Montana, Missoula, MT 59812.

10. Cowan Schwartz, R. (1983). *More Work for Mother.* Basic Books, New York.

*Technology has changed what housewives do but not how long it takes; thorough scholarship and lucid writing.*

11. Foster, G. M. (1962). *Traditional Cultures and the Impact of Technological Change.* Harper & Row, New York.

*An anthropologist's perspective. The most useful chapters are on the barriers to change — cultural, social, and psychological — and the stimulants to change.*

12. Fradkin, E. M. (1989, May). Two lives for the Ariall. *Nat. History*, 38–49. Quotation on page 46.

*Fradkin is really interested in how people can survive periods of famine. Nomadic people actually do it well but their life is hard. The article compares life in a mission station with life in arid grazing areas.*

13. Pacey, A. (1983). *The Culture of Technology.* MIT Press, Cambridge.

*Good place to begin reading about nontechnical issues. Considers technological innovation, control of technological change, the satisfactions of technology, domination over nature, differential impact on women, and designers versus users in detail, with much insight.*

14. Reardon, K. K. (1984, September/October). Its the thought that counts. *Harvard Business Rev.* **62**(5), 136–141.

*Fun to read article on culture in different countries, specifically related to gift giving.*

15. Roy, R. (1985, March). User needs and Appropriate Technology. *Appropriate Technol.* **11**(4), 7–8.

    *Describes the acceptances of improved cook stoves in Indian and Zimbabwean villages and of new technologies in general.*

16. Saito, K., and Weideman, C. J. (1990). *Agricultural Extension for Women Farmers in Africa*, World Bank Discussion Paper No. 103. World Bank, Washington, DC.

    *Suggestions on how to provide agricultural extension services to women; many case studies.*

17. Skjonsberg, E. (1989). *Change in an African Village*; *Kefa Speaks*. Kumarian Press, West Hartford, CT.

    *The author spent 16 months in a Zambian village, Kefa, and reports on life there, focusing on women.*

18. Smith, J. (1980). Something old, something new, something borrowed, something due. In *Women and Appropriate Technology*. Women and Technology Project, 315 S. 4 Street, Missoula, MT 59801.

    *Thirty-page booklet, mostly about the United States and why women tended to be left out of the Appropriate Technology movement; contains an introduction to Appropriate Technology with a directory.*

19. UNICEF (1983, December). *A Practical Guide to Improve Fish Smoking in West Africa*. UNICEF, New York.

    *Sample of training manual used in Third World; describes construction, cost, uses, and so forth; many pictures.*

20. Wiser, W., and Wiser, C. (1963). *Behind Mud Walls 1930–1960*. Univ. of California Press, Berkeley.

    *The Wisers were missionaries to a village in India who learned to become anthropologists and both ministered to the village and studied it. Mrs. Wiser returned in 1960 and to see how it all turned out. A real pleasure to read.*

# PROBLEMS

1. It has been asserted that the cultural institution of marriage is threatened by technological changes, such as kitchen appliances making food preparation more feasible for a single person, new types of jobs making it possible for women to be financially independent of men, and contraception. Argue either for or against the assertion.

2. Another assertion is that improvements in transportation — mainly commuter trains and high-speed highways — have doomed cities and thus the cultural life (opera, theater, and art exhibitions) that have traditionally flourished in cities. Argue for or against the assertion.

3. It seems technologically feasible for most office workers to do most of their work at home (telecommute) connected to other workers by high-speed data links. How would cultural attitudes promote or work against the institution of such technology?

4. Suggest a plan for the World Bank to decrease deforestation for firewood.

5. Describe a technological change meant to help an entire community that ended up helping only a subgroup — a subgroup that was better off to begin with. (Suggestion: use of methane digesters in a village, use of computers in primary schools as optional enrichment, and snowmobile trails in wilderness areas).

6. In the United States, some jobs became primarily associated with women and some with men. Why might this have happened? What can be done about it?

7. Some people fear that a move to simpler technologies in the industrialized world will result in ''back to the kitchen'' for women. In what ways does this fear seem valid or invalid?

8. Is not a community better off if a mechanized factory can produce things (e.g., textiles) cheaper than they could be made at home? Explain.

9. What are the risks and benefits to a household in diverting cash from domestic needs to a new business?

10. Is it likely that cash directed to women in the household will really have a more beneficial effect on the community than cash directed to men?

11. Some contemporary aid projects have a stated purpose of directing cash to women, which is not the norm. What are the risks of doing something like this, contrary to the local culture? What can be done to lessen the risks?

12. When Japanese factories opened in the United States, much time and effort was spent at meetings with workers. In terms of the ideas of this chapter, what were the purposes of these meetings?

13. How would you have tried to meet the needs of the Rendille people?

# *16*

· · · · · · · · · ·

# TECHNOLOGY POLICY

· · · · · · · · · · · · · · · · · ·

---

## Soviets now setting the pace in space

Their low-tech program succeeds while U.S. efforts remain grounded

... It is widely agreed that Soviet space technology is no match for that of the United States in terms of sophistication — in microelectronic circuitry, for example, and in the life span and "smartness" of its space-flight systems.

And yet the Soviets' low-tech space program is more resiliant than that of the United States. They have outpaced this country in sheer experience on several fronts in space, and are poised to do so in others, according to a wide sampling of space experts.

---

**Reproduced by permission of Providence *Journal*, Monday, July 20, 1987.**

## WHY WE STUDY POLICY

We study how policy is made and implemented by governments, and other organizations, to answer several questions: In a specific situation, does Appropriate Technology makes sense? If it does make sense, what must be done to make it succeed [3]? What are the obstacles to its acceptance? How can organizations be mobilized to foster the acceptance of Appropriate Technology?

The reason for examining policy making is that widespread use of Appropriate Technology may not come without policy decisions. A person who thinks Appropriate Technology can be beneficial needs to understand the process of establishing policies so as to contribute when decisions are made. In this chapter, we provide a description and some examples of policies, examine how to overcome obstacles to Appropriate Technology, and examine the policy-making process. An example of how energy policy might be set in Malawi will clarify this process. Finally, we consider what mechanisms government agencies have to implement policy, particularly an Appropriate Technology policy.

## WHAT IS POLICY?

We use *policy* to mean a consistent combination of objectives, plans, and programs. *Objectives* are what one wants to do, *plans* are how one will do it, and *programs* are what one does. *Consistency* means that all these programs can be done simultaneously, and that success in one does not preclude success in another. A policy is inconsistent if it distributes a limited amount of money among so many programs that none receive sufficient money to accomplish goals or if a program of one kind counters or diminishes the effects of a program of another kind; for example, a program of giving food to Third World nations may be inconsistent with a program of encouraging local agriculture.

An illustration of a policy, its objective, plan, and programs was President Carter's proposal to accelerate the use of solar energy [4,8]. One objective was to supply 20% of U.S. energy demands in the Year 2000 from solar sources — trees and grains were included as solar sources. The plan was made up of several programs, among them tax credits for wood-burning stoves, tax exemption for gasohol, and interest subsidies for installers of solar equipment. The programs were, of course, specified in detail — how much credit or exemption, who would inform potential beneficiaries, when the exemptions ended, and much more. The policy was never implemented so we do not know if the programs were consistent — whether the nation could have afforded the costs of all the programs together and whether (a far-fetched possibility) the smoke from wood-burning stoves would have interfered with solar heaters.

A contrast in policies is demonstrated by India's and China's methods of industrial development in the 1960s. India's government decided to invest in large-scale industry, competitive with European, Japanese, and American companies, emphasizing steel production. The country experienced a rapid growth in heavy industry with great expansion in production capacity. China, on the other hand, followed Mao Tse-Tung's "walking on two legs" policy in which heavy industry was one leg and agriculture, with light industry, the other. The rationale was that both legs were needed, neither shorter, to achieve fastest economic growth.

The outcomes of these two policies were quite different. With India's strong focus on industrial growth, agricultural development and rural farmers were neglected. India placed much emphasis on increasing production of industrial goods rather than examining the demand. Consequently, products of the heavy industry could not always find a market and some factories have never worked at full capacity. The Indian semiconductor industry, for example, spent $45 million in one year to produce devices worth $15 million.

In China, industrial development policy had several components. One component was to support small local industries in the rural areas, in smaller cities, and even in some villages. Development in rural areas was encouraged by giving authority to officials in these areas to attract small factories. The policy was successful and local industries have increased both the variety and the quality of products available to consumers. Another component of the policy recognized agricultural development as the foundation of industrial development. Agricultural development requires agricultural investments, such as farm machinery and fertilizer, so the policy, being consistent, included the support of factories manufacturing machinery and chemicals. The new businesses were also given inducements to locate away from the large cities. The result of the policy was widespread growth, benefiting a large part of the population, of both food and consumer goods.

Actually, India changed its policy in the 1970s. It is now doing more for its rural areas. Another component of its current policy aims to make it an industrial leader to the less industrialized countries rather than replicating Western industry. As part of this policy, India has been selling to the Third World smaller and simpler machines than have other exporting nations.

## WHY MAKING A POLICY IS USEFUL

Policy making is done by most companies, governments, and other organizations. Why do they bother? To make sure they are doing the things that will most benefit the organization and to make sure they are doing those things effectively. Peter Drucker, a writer on management, is supposed to have said: "It is better to do the right thing than to do things right" [2]. Of course, it is best to have a policy which does both—one in which the organization works on its most important problems and does that work well.

Japan's remarkable industrial leadership is often attributed to its policy of targeting particular industries—steel, automobiles, consumer electronics, and computers. Engineering efforts were concentrated on these target industries, often through cooperative research. This targeting policy results in products being brought into production quickly because people are working on the same problem but not duplicating each others' efforts. An essential part of the policy is choosing the right industry, one which will be important in the future, one which Japan can dominate, and one which will lead to other significant industries. Although this example has been over-simplified, one can see how a policy both steers industries into worthwhile areas and generates results within those areas.

Policy making improves an organization in another way: It focuses the attention of people in the organization. It ensures that everyone involved knows what the organization is doing. It can give people in the organization an opportunity to participate in planning, and participation often leads to commitment. Without an explicit policy, understood throughout the organization, a risk exists that people will work at cross-purposes. To be specific, if a foreign aid organization has a stated policy of doing Appropriate Technology, then it is unlikely that a project manager in the field will chose a high-technology machine, especially if she or he has been involved in the decision to use Appropriate Technology.

# OBSTACLES TO INSTITUTING APPROPRIATE TECHNOLOGY

Understanding policy making can help in overcoming the obstacles to general employment of Appropriate Technology. The important obstacles seem to be of three kinds: material, organizational, and attitudinal. Appropriate Technology may not produce enough or do enough to meet people's needs. We will call this a *material* obstacle. Appropriate Technology may not match the way a company or agency is organized. This obstacle we will call *organizational*. Also, an obstacle derives from people's attitudes and will be called *attitudinal*. We will discuss in turn how to overcome each of these obstacles.

## *Overcoming Material Obstacles*

The basic material concern is whether Appropriate Technology can do the job. Potential users may believe that small-scale or unconventional solutions are insufficient or risky, and they may be correct. Furthermore, a material aim of a technology policy may be more than supplying food, shelter, or energy. It may be to develop technical expertise that will lead to a fully industrialized economy. Planners may doubt that Appropriate Technology provides the expertise that a modern nation needs. Will a nation using Appropriate Technology remain poor because the technology does not produce enough and/or because the technology training leads up a blind alley?

This obstacle, dealing with the outputs of a particular technology choice, is especially serious for Appropriate Technology because of the difficulty in predicting how well Appropriate Technology will work in a different setting. Appropriate Technology is site specific and so designs which work at one place may not transfer effectively to another. The obstacle is that not enough is known, in many cases, about Appropriate Technology to transfer it reliably. This relative lack of experience may make it harder to predict ahead of time whether Appropriate Technology, compared to more familiar technologies, will actually meet requirements.

Examining material outputs and inputs is an essential part of policy formulation. A valid way of making policy deals explicitly with the outputs, material things (e.g., food and energy), of a particular policy choice. Policymaking is a systematic analysis of costs, benefits, and risks. A primary question it answers is what will be the result of, in this case, a choice of Appropriate Technology. Needless to say, Appropriate Technology may not be the best way to meet material needs in every situation. The analysis should show if it is an effective way and thus attempts to use Appropriate Technology when it cannot produce enough can be avoided. One should not, however, expect too much from an analysis. Not all outputs of a policy decision can be predicted. At the least, though, thinking carefully about whether Appropriate Technology should be chosen will indicate the risks and the likelihood of success. The process of making policy should reduce the risk that Appropriate Technology is used in situations in which it cannot meet material needs.

Effectively developing a policy can ensure that other objectives, such as training of workers, are reached. If an objective of a technology policy is to prepare a base for industrial expansion, then when the policy is being developed a subobjective, encompassing the preparation, needs to be recognized. Policy making includes listing all the objectives of the policy and integrating them into the final policy. One cannot expect that it will be possible to meet all objectives but the result of explicitly determining policy should show how one objective was traded against another.

## *Overcoming Organizational Obstacles*

Organizing an Appropriate Technology project is usually entirely different from organizing other technology projects. Appropriate Technology, in most actual cases, consists of many people working on small projects, each project specially matched to local needs and resources. A shelter project, for example, might be organized into many autonomous teams, each composed of a few people, making bricks with local soil and building houses consistent with local customs and particular needs. In contrast, a high-rise apartment building project would be organized to put decision making in the hands of a few supervisors. Neither industry nor government, the two sources of most of the world's management experience, tend to implement projects requiring primarily local decision making, so little is known about how to set up this kind of organization. How would you set up a program to install windmills on all the medium-sized farms in the United States when wind conditions, electricity needs, and farmers' finances vary so much throughout the country? In such a project, how much supervision from the top can be given to field workers? How can their success be monitored from the central office? These questions are difficult and few successful models are available to guide a planner.

Another organizational obstacle is that labor-intensive technology may be unattractive to managers. It is nearly always easier to manage machines than people. For example, compare the effort required to get a small office computer to do its work (once the programs are installed) with the effort required to manage an office staff performing the same job. Running a four-person office is more difficult, most supervisors would agree, than using a small computer. Even in the Third World, where labor costs are low, managers may prefer to use mechanized production so as not have to deal with a large labor force with unfamiliar customs and demands. The added jobs brought about by Appropriate Technology may benefit a nation but be a burden to the manager involved.

Appropriate Technology expects much not only of managers but also of the people who operate it. It takes more labor and thought to grow crops organically than to apply herbicides and pesticides. Construction of a well-insulated house requires more care than of a noninsulated one. A wind-powered generator needs attention during stormy weather. Some Appropriate Technology devices, such as urban gardens or community methane digesters, require cooperative effort which is often more difficult than simply having one person give orders. An organization will not work if too much is demanded of people within it. The challenge is to organize an Appropriate Technology project so individuals are given realistic tasks.

An organizational obstacle for Appropriate Technology is that most technology transfer to the Third World has been by multinational companies, organizations which are often biased against Appropriate Technology. In many organizations people tend to think alike and use the same approach to problems—the approach which has been successful in the past. In many multinationals the bias has been toward high technology, probably because their historical competitive advantage in the United States, Japan, or Europe has been in high technology. This bias is reflected in the technical specifications which companies and development agencies normally have that set the standard to which their organization will build. The obstacle to Appropriate Technology is that these technical specifications tend to be the same as those that the multinational uses at its home, even though such factors as raw materials and uses of manufactured products are different in the Third World. Ingrained practices within the organizations currently performing most technological transfer act against Appropriate Technology.

Valid policy making should either develop an organization that overcomes the obstacles just described or conclude that Appropriate Technology does not make sense in the particular case. Deciding how to implement a policy is part of deciding on the policy. Making policy forces planners to specify responsibilities for both managers and technicians, exposing those positions in the organization with unrealistic duties. If policy is determined systematically biases should become apparent and can be compensated for.

Appropriate Technology tends to consist of bottom-up programs, in which the significant decisions and ideas come from the bottom — the field. When determining policy in a specific situation the policy maker needs to consider how to organize the effort so the bottom-up process works. If no way of organizing seems to be possible, then the Appropriate Technology approach should be discarded. Going through the policy-making exercise should tell the designer in advance if the organizational obstacles can be overcome.

### *Overcoming Attitudinal Obstacles*

One basic attitudinal obstacle to Appropriate Technology is sometimes called the allure of high technology — the attraction of what is new, complicated, and mystifying, and that has as many indicator lights and knobs as possible. Something simple and clear is perceived as inferior. To some extent this attitude seems to be universal; even designers committed to Appropriate Technology can refine and improve an originally simple device into something complex, taking it out of the Appropriate Technology sphere entirely. Solar heating for domestic use is an area in which things can often get out of hand, with microcomputer controls and motorized collectors to track the sun. The obstacle is focusing on the sophistication of the device rather than on how well it meets needs. The tendency to overcomplicate will also be discussed in Chapter 17.

Another attitudinal obstacle is that Appropriate Technology is just plain unfamiliar. Tractor manufacturers have made sure that everybody who cares knows about fancy, high-powered, high-technology farm machinery, but few people push simple but functional machines. This unfamiliarity obstacle relates to the material and organizational obstacles. Because government leaders do not believe that Appropriate Technology can actually meet material needs they are nervous about trying it and thus do not consider it seriously. These same leaders may not believe an Appropriate Technology project can be effectively managed and so rule out the approach from the start.

An obstacle in the Third World is that some national leaders seem to prefer high technology (although other national leaders have embraced Appropriate Technology). One reason suggested is that modern factories are considered a badge of honor, a symbol of how far the country has progressed. Some Third World governments may believe the Appropriate Technology movement is a way of putting Third World countries down — of declaring that they are not prepared for high technology. Another interpretation of why Appropriate Technology is pushed on Third World countries is that it keeps them from becoming industrial rivals. Even if these attitudes are not true, they still hinder the acceptance of Appropriate Technology.

The attitudinal obstacles are basically that Appropriate Technology is not taken seriously or not considered at all. If a policy-making study results in a recommendation for Appropriate Technology, then that technology will gain a needed legitimacy. If Appropriate Technology is treated as a serious contender in deliberations, then people will consider it regularly when planning policy. The process of formulating policy can improve the image of Appropriate Technology, thus overcoming obstacles to its acceptance.

## THE PROCESS OF POLICY MAKING

Figure 16.1 shows a model for policy making; a more complete description is given in Ref. [11]. The figure is misleading because it implies that each stage is done separately, for example, that one can select the optimum policy before considering implementation. In reality, each stage must be considered in relation to each of the other stages; both external and internal factors should be taken into consideration before choosing objectives if the objectives are to be sensible. The model of Fig. 16.1 is useful, however, because it helps the planner break down the problem into manageable pieces. It is also useful because it reminds the planner of each of the necessary steps.

Now we discuss the blocks in Fig. 16.1 individually. *Determining objectives* can be the most fruitful step in the whole policy-making exercise. It has been said that if you do not know where you are going, any road will take you but you will not know when you have arrived. A clear statement of purpose will help people actually carrying out the project decide how to proceed if something unexpected happens. It is usually best to make objectives definite, with a time duration specified; for example, "70% of the villages in this province will have VIP latrines at the end of 18 months." If objectives are definite, a manager is able to recognize if the project is on schedule or if it accomplished what it was intended to do.

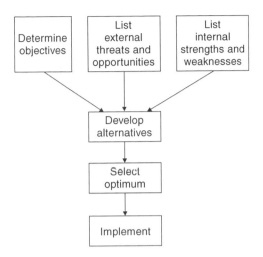

**Fig. 16.1. Model for making policy.**

However, not all objectives are quantifiable: How does one measure the "quality of life in the villages"? Worse is the perversion of a hard to measure objective by use of an easily measured index. Trying to measure improvement in village life, for example, by counting the number of visits made by social workers encourages short, perfunctory visits, solving mostly the easy problems. In any case, objectives need to be realistic in terms of both the resources available and the actual situation: What really can be accomplished in this district with the staff likely to be available? Realistic objectives motivate people. Unrealistic objectives are often ignored.

*External threats and opportunities* are the things the planners do not have under their control. The words "threat" and "opportunity" are from the point of view of the organization. An opportunity is an opportunity for the organization to do something; for example, to improve health care in a country. A threat to a nongovernmental organization might be a new government regulation that no more than 50% of the staff be expatriates. Note that an opportunity for an organization may actually be a threat to the country: Deforestation is a threat to a nation but an opportunity to an organization that has expertise in fast-growing trees. Similarly, an organization can turn a threat into an opportunity. If a new regulation states that 50% of the staff must be local people then resources for training may be easier to mobilize: The organization can use the new government regulation as an opportunity to start a training program. Not all opportunities appear as unmet needs. A technology new to a region can be an opportunity. High-yielding seeds represent a real opportunity to improve agriculture in a region where they are not currently used. Another external factor is the policies of other organizations. If one does not introduce new seeds into a particular country, will another group do it? Turf battles should be avoided (pardon the pun).

*Internal strengths and weaknesses* are usually people, money, or facilities. People here usually means expertise: What does our organization know well or do well? In what areas do they lack experience? In nearly all cases, it is difficult to change an organization's basic area of competence, as big corporations that diversified unsuccessfully into many unrelated businesses in the 1960s discovered. If a proposed policy does not match the skills and backgrounds of the people in the organization, one should be wary. The second aspect of internal strength and weaknesses is money. Calculations of financial resources and monetary needs are usually straightforward, except that potentially available money — that which can be raised — should be included within resources in addition to the cash already on hand. If we are deciding policy for a company, money which can be raised is generally that which could be borrowed. If the organization gets much of its money from contributions, then in estimating how much can be raised one must account for the different donor appeal of different projects. Some needs will be more fashionable than others at a particular time and will attract larger gifts. Facilities, the third aspect of internal factors, are sometimes important. A laboratory and workshop might be necessary if one wanted to design a high-performance water turbine. A well-equipped but specialized workshop could be a weakness if it encourages only approaches consistent with its specialty. In practice, it is seldom fruitful to make a major effort to distinguish between internal strengths and weaknesses, but it is important to realize what the organization can do well and what it cannot do well.

It is undoubtedly obvious what is done in the stages labeled "develop alternatives" and "select optimum" in Fig. 16.1. Doing these things in practice though is not always straightforward. We will discuss these stages after we provide an example.

The final stage is *implementation* — making it work. Implementation is usually done by breaking a project into subprojects small enough to be understood and handled. Each subproject can then be made the responsibility of a person or small group which actually does the work. The person or the group lays out the work plan in detail, constructs schedules and budgets, and devises ways of monitoring progress. The subproject schedules and budgets must be merged into an overall schedule and budgets and an overall monitoring scheme must be devised to catch problems, e.g., subprojects falling behind schedule or spending too much money.

Subprojects should be made autonomous, that is, able as far as possible to function independently. Autonomy builds commitment; it ensures that designs meet local needs. It does create risks, however. One risk is that something will go wrong in a subproject and because no one else knows of the difficulty no one assists. Ways of monitoring that do not stifle initiative must be found. A second risk is that good ideas (e.g., innovative designs) developed in one subproject may not become known in other subprojects. Effective means of transferring information must be found. Some electronics companies, faced with the same problem, hold a party every Friday afternoon at which designers get together and talk about their work. Appropriate Technology functions well when people are fully involved and working on problems specific to their needs. Sharing of successes and mistakes can be useful. The gains from sharing must be balanced against the losses from a pressure to conform and from the distraction of another meeting.

Job assignments must also be made. Jobs must be big enough, that is, have sufficient challenge and opportunity for learning, to be satisfying but they must be small enough to be mastered. It should also be feasible to evaluate each person's work. It is important throughout the world, but especially in the Third World, that job assignments be made in a way that fosters the professional growth of people in the organization. People grow through new jobs and stronger challenges. A fundamental objective in the Third World, of course, is to have local people eventually take over the project. This objective must be kept in mind when jobs are being planned, in fact, through all phases of implementation.

## EXAMPLE OF POLICY MAKING

The example which follows demonstrates how an energy ministry in a Third World country might determine policy. Specifically, we consider how the nation of Malawi can meet its energy needs. This example comes from a report written in 1981 [3]. The data used are from the late 1970s. We follow the plan in Fig. 16.1.

The objectives of the policy are determined in consultation with other government ministries. We use the following objectives:

*1.* Maintenance or enhancement of the supply of cooking and heating fuel in the villages

*2.* Preservation of a dependable supply of vehicle fuel

*3.* Reduction of foreign exchange expenditures

Pertinent external factors are the price of oil on the world market, the undependability of the rail link to the coast through Mozambique (anti-government guerrillas there have destroyed bridges), and the possibility of a large hydroelectric plant at Cabora Bassa on the Zambesi River in nearby Mozambique, which could supply power to Malawi. An externally based opportunity is the development of alternative energy technologies, e.g., ethanol as a substitute for gasoline. Much research on ethanol has been done in Brazil and South Africa. Another opportunity to the energy ministry is the possibility of receiving foreign aid to improve the railroad inside Malawi.

The major internal factors are the sources and uses of energy, shown in Table 16.1 and Table 16.2, respectively. Table 16.1 shows that by far the greatest source of energy is wood, which is chiefly used either for process heat in the tobacco and textile industries or for cooking in villages. (Blue gum is a tree, actually eucalyptus, commonly grown on tree plantations.) Some observations will make Table 16.1 more meaningful. In 1978, the area of government wood plantations was 68 kha. Also in 1978, about 1700 kha in Malawi was under cultivation and about 800 kha of usable land was not cultivated. Unused hydroelectric power is also available. Planned or existing installations have a capacity of 450 megawatts (MW), corresponding to approximately $3.9 \times 10^6$ MW-hr. Of this 450 MW potential about 244 MW should be installed by 1995. The unused capacity of the Cabora Bassa–Zambesi installation in Mozambique is much larger, which may inhibit the installation of hydroelectric plants in Malawi.

The objectives of an energy policy would be, to a large extent, met by replacing the imported sources in the left-hand columns — diesel fuel, petrol (i.e., gasoline), and coal — by the renewable sources in the right-hand columns — wood and particularly hydroelectricity.

Another internal factor is how much foreign exchange is spent on energy. About 12% of the total value of imports is fuel — 15% if fertilizer, often from petrochemicals, is included. Diesel fuel, in fact, used more

#### Table 16.1. Sources of Energy in Malawi and Percentage of Energy Content

| Source | Percentage of energy content | Annual amount |
|---|---|---|
| Wood | 94.0 | Equivalent to approximately 80 kha of blue gum |
| Diesel fuel | 2.5 | Approximately 87 megaliters (Ml) |
| Petrol (gasoline) | 2.5 | Approximately 57 Ml |
| Electricity | 0.9 | Approximately $380 \times 10^3$ MW-hr |
| Coal | 0.8 | Approximately 34 kilotons |
| Paraffin (kerosene) | 0.3 | Approximately 10 Ml |

## Table 16.2. Energy Use In Malawi

| | Diesel fuel (Ml) | Petrol (Ml) | Coal (kilotons) | Wood (kha) | Electricity (MW-hr) |
|---|---|---|---|---|---|
| Transportation | | | | | |
| Private cars | | 57 | | | |
| Commercial vehicles | 14 | | | | |
| Railways | 6 | | | | |
| U.T.M. limited[a] | 5 | | | | |
| Industrial | | | | | |
| Tobacco | | | 5 | 32 | 4 |
| Tea | | | 6 | 4 | 40 |
| Sugar | | | | | 130,000 |
| Textiles | | | 12 | | 30,000 |
| Other: | | | | | |
| Construction equipment | 36 | | 10.5 | | 124,000 |
| Nontransport | 16 | | | | |
| Farmers | 8 | | | | |
| Domestic | | | | | |
| All homes | | | | 44 | 50 |

[a] The bus company.

foreign exchange than any other commodity coming into Malawi in 1979 and gasoline used the second largest amount.

The next stage in the policy-making process of Fig. 16.1 is to develop a set of alternative projects. This could be done by observing which experimental projects have been successful in Malawi and learning which energy technologies have worked in other countries. The set of alternative projects developed is shown in Table 16.3. For each project, costs and benefits can be estimated using the techniques described in Chapters 2–12. These costs and benefits are also shown in Table 16.3. (The currency unit is the Kwacha, abbreviated K, equal to about $1.20 in 1979.) The ethanol strategy listed is based on the production of alcohol from sugar. To grow sufficient sugar cane would take 73 million ha. The areas of blue gum shown are for sustainable yield, that is, assuming trees are harvested after 8 years of growth.

Additional factors need to be considered besides those shown in Table 16.3. One factor is lifestyle benefits, for example, rural electrification will benefit villages in many more ways than simply reducing deforestation. An

**Table 16.3. Energy Strategies, with Costs and Benefits**

| Strategy | Cost | Benefit |
|---|---|---|
| Biogas digesters | $K40 \times 10^6$ | Replaces $K23 \times 10^6$/ year diesel fuel |
| Railway electrification | $K56 \times 10^6$ | Replaces $K1.5 \times 10^6$/ year diesel fuel |
| Use of ethanol as motor fuel | $K44 \times 10^3$ | Replaces $K15 \times 10^6$/ year petrol |
| | $73 \times 10^3$ ha | |
| Replacement of small diesel engines by wind or hydropower | $K4 \times 10^6$ | Replaces $K1 \times 10^6$/ year diesel fuel |
| Conservation | | |
|   Efficient village stoves | $K3.5 \times 10^6$ | Replaces $176 \times 10^3$ ha* blue gum |
|   Industrial energy conservation | $K1.2 \times 10^6$ | Replaces $34 \times 10^3$ ha* blue gum |
|   Burn agricultural waste in factories | $K3.1 \times 10^6$ | Replaces $2.7 \times 10^3$ ha* blue gum |
|   Electrify homes in rural areas | $K405 \times 10^6$ | Replaces $352 \times 10^3$ ha* of blue gum |

* Area required for sustainable yield.

intangible factor is administration. It will be much easier, for example, to persuade the management of Malawi Railways to electrify than to persuade the 2 million heads of households in Malawi to use an efficient woodstove or 100,000 cattle owners to collect manure for digesters.

Selection from the projects listed is done on the basis of both the data in Table 16.3 and intangible factors, including environmental and cultural concerns. Some person or group has to weigh the relative importance of the costs, for example, of the land to be set aside for tree plantations against the use of foreign exchange for purchasing electric water pumps in villages. Table 16.3 shows, for example, that an efficient village stove would be of greater value than using agricultural wastes. Widespread implementation of biogas digesters would also be valuable but expensive, whereas replacement of small diesel engines by wind or hydro would be relatively inexpensive but not as worthwhile. A sensible choice for the energy ministry in Malawi might be an efficient village stove project and a replacement of small diesel engines project.

Once a project is chosen, it must be implemented. For example, if the efficient village stove project is to be carried out, someone must be made responsible. This person must decide what problems must be solved: Is more design work needed? How much training of local people is required? What is the best way to inform potential users about the stove? and so forth. Then a group of people with the needed skills can be put together. These people can decide what specific tasks have to be done and set a schedule. They will also have to consider ways of monitoring the projects and, finally, they must go to work and do it.

## *Development of Alternatives*

The least routine stage but probably the one that affects the outcome the most is the *development of alternatives*. This is where the inventing is done, although many good alternatives are not fresh inventions but adaptations of something successful somewhere else. It is unlikely, for example, that any of the alternatives of Table 16.3 will require major inventions. As noted in that example, ideas for alternative policies can be gotten both inside and outside the organization.

How do we nurture innovations within an organization and how can research or development be managed? In the Malawi case, how does the government encourage people to design an improved woodstove or experiment with small-scale hydroelectric systems? Within a company or aid organization, how does management ensure a flow of new ideas or new products? When something new is being designed, the manager usually knows less about the really difficult design questions than the innovator. The manager cannot help the innovator with technical problems or even judge if a good job is being done. The manager is really hostage to the ability and commitment of the person doing the work. Certainly the history of innovations shows that "benign neglect"—letting engineers experiment and design with little supervision—has produced more beneficial results than tighter control, at least in the initial stages of new product development. Management should focus their efforts on evaluating projects brought to them rather than trying to control the design of those projects.

In order to see whether there is more that managers can usefully do to nurture innovations we might examine the way innovations take place [10]. The process basically has three steps, as shown in Fig. 16.2. In the first stage, *experimentation*, the innovator tries a new idea: a farmer in Malawi might read of a methane digester and build one on her/his own. In the second stage, *learning and adaptation*, the idea is duplicated by other

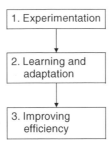

**Fig. 16.2. Stages of innovation.**

farmers. These other farmers learn more about how digesters function in the Malawi climate using Malawian wastes — the design of the digester is adapted to meet the Malawian situation. In other organizations analogous events occur: Different users try the innovation, learn what it can do, and adapt it to their own needs. In the third stage consideration is given to where digesters can be used most efficiently and the design is made very specific for that situation. In general, the third stage begins when experimentation is over: The purpose here is to fine-tune the innovation for specific applications and make life easy for the user.

The role of the manager, that is, the person responsible for new technology, depends on the stage of the innovation. In the first stage, the prime need is to encourage a wide range of experiments — "let many flowers bloom!" Experimentation can be encouraged by giving money for supplies, by having machine shops or similar facilities available, by giving suggestions, and by providing catalogs and magazines describing work done in other countries. The worst thing that a manager can do at the first stage is kill a good idea. In the second stage, the prime need is marketing — encouraging users to try an unproved innovation. The worst thing a manager can do here is miss a potentially worthwhile application. In the third stage, the prime need is control — to stop making refinements so users can actually apply the new technology. Refinement can and should be made in parallel with use but the manager needs to protect the user from changes in the working device or system. A succession of changes in an installed solar heating system, for example, is a major nuisance to the home owner.

These roles of a manager apply straightforwardly in the development of an Appropriate Technology innovation. The manager must encourage experimenters, tinkerers, and inventors or at least not thwart them. Evidence exists that too much support, i.e., money, can distract designers at this stage: Leanness has its virtue. Besides encouraging the people to experiment, the person responsible for new technology should also be beating the bushes for inventions and ideas that have worked, getting others to try them, and taking good ideas from one place to another— corresponding to the second stage in Fig. 16.2. The person responsible should also decide when a technology is mature enough to be distributed widely to those users who will not want to make modifications; in other words, decide when the design should be frozen and widely released.

The other source of innovations, besides inventions made within the organization, is transfer from outside. Technology transfer, especially to the Third World, is another important subject about which little seems to be known. Engineers get some of their ideas from reading journals, catalogs, and internal company reports (if they are employed by a large company). They seem to get more ideas from listening to other engineers on the job or at professional conferences, company seminars, or less formal meetings. It appears that even at professional conferences the informal discussions are when most of the information is transferred. Vendors (i.e., salespersons) are a major source of information to many companies but, of course, are not always disinterested—obvious risks exist in depending on salespeople if the buyer is technically naive.

Universities are an important vehicle for technological transfer; faculty members generate some new technology and are paid to keep up with the rest. Undergraduate training, of course, takes place just once in an engineer's life but graduate programs or short courses bring participants up to date later in their careers. Faculty members give seminars off campus and consult. Both activities bring new ideas into organizations. If a person consults for more than one organization he or she may transfer information between organizations.

Engineers in the Third World are at a disadvantage in terms of technology transfer. Consider each of the modes just described. Third World engineers have fewer colleagues to talk with. Journals and catalogs are often harder to find. The university system is less elaborate. There are fewer vendors so it is harder to compare claims. Trips to professional conferences are less likely. Similarly, Appropriate Technology, as a field, tends to be at a disadvantage—fewer practitioners, fewer vendors, fewer journals, fewer universities, and fewer conferences. Few companies use Appropriate Technology often, so the reports or meetings that transfer technology within most companies do not usually describe Appropriate Technology. The implication is that managers of Appropriate Technology programs in the Third World need to be especially concerned about getting up-to-date information to their staffs. In the past few years, the situation appears to have improved, with more directories of manufacturers available, more (but still few) university courses, and more conferences. In making choices of technology, a manager should realize the bias against Appropriate Technology in dissemination of information and compensate for this bias.

## *Selecting Optimum Policy*

A basic problem in deciding what to do (the *select optimum* box in Fig. 16.2) is trading off tangible factors with nonquantitative ones. How many dollars is a clean environment worth? Is it better not to have to carry water a half mile from a village well four times a day or to have a midwife present at every birth? Do community gardens or storm windows accomplish more for an inner-city home? How does one tell? Convincing ways of answering these questions are not known. Straightforward quantitative approaches have been described. An example [4] is computation of the cost–benefit trade-offs of reducing air pollution in the Los Angeles area [4]. The benefits are measured by decreased health care costs and the monetary benefits of increased visibility. The costs are measured by the capital and operating costs of removing sulfur from gasoline. Such an analysis seems to leave out so many important factors, such as maintenance of buildings, plant life, illness due to air pollutants, and so forth, that it is difficult to find the conclusions compelling. The best advice one can give about how to weigh different kinds of costs and benefits is simply to use good judgment.

Because judgment is required in making the choice, the biases of the decision makers are more important than they would be if the data determined the answer. Appropriate Technology can, again, be at a disadvantage in situations in which personal biases are important if the policy planners have had extensive background in other kinds of technology. People naturally use the approach they are most familiar with; as someone said, "If you are accustomed to using a hammer, every problem looks like a nail." In order for a valid choice of technologies to be made, it is important that these biases be recognized, either by ensuring a variety of backgrounds within the group making the choice or by consciously compensating for possible predispositions.

When choices are made in large organizations, each member of the group making the choice usually represents a constituency. The concerns of these constituencies are inevitably reflected in the choice. Compromise between the desires of various constituencies is an obvious and essential part of government policy making but the same attitudes of protecting one's friends can be present also in large companies or nonprofit organizations. In a typical aid program, for example, groups will be working on food, health care, small-scale industries, and so forth. If the choice of projects is made by a committee, then committee members' votes may reflect the desires of their constituents as well as their own beliefs. Needs that do not have an organized constituency may fare poorly in the voting.

The national constituency for Appropriate Technology is not as organized as that of many competing approaches. As noted, mostly high technology is produced by large corporations, which are experienced at working through the political process (e.g., the electric utility companies, for which government regulation is a primary aspect of doing business). Appropriate Technology comes mostly from small organizations or individuals not linked together, who often lack the experience and funds to influence decision makers. While most people approve generally of Appropriate Technology, only a few influential people support it passionately. To get lawmakers' attention, passionate support by influential people often is needed. A necessity if Appropriate Technology is to be considered equally with other approaches is that it develop an organized constituency — a lobby pushing for it. A book, *The Solar Energy Transition* [9], argues strongly for a solar lobby if solar energy is to be widely accepted. The same arguments apply to other aspects of Appropriate Technology; an organized lobby may be required to get them accepted nationally.

We have discussed choice situations in which many costs and benefits cannot be expressed in terms of dollars, so intuition and biases are important in the decision making. In some cases, of course, major costs and benefits of alternative policies can be measured quantitatively. When this is true, mathematical methods can be used to select the best policy or the optimum combinations of policies [8]. Realistic optimization problems require computers for solutions. Computers are becoming cheaper and effective planning programs which run on personal computers are now available [6]. Even if not all parameters of a problem can be quantified, the available data can suggest promising approaches to a design.

## TACTICS FOR IMPLEMENTING A POLICY BY A GOVERNMENT AGENCY

Once a policy has been determined it must be implemented. In the Malawi example, efficient stoves must be designed and built and they must be put to use. Government agencies seldom achieve success by simply announcing a policy. How could the Department of Energy insulate every house in the United States, publicize the value of insulation, try to actually do the required installation, and try to inspect every house and fine noncompliers? None of these approaches seem realistic. Nearly always the implementation of a government policy must be indirect. The tactics that can be used for implementation fit into five categories: operations, information/assistance, administration, finance, and standards. These are listed in Table 16.4, which is based on Ref. [5].

Some comments on Table 16.4 may be useful. *Operations* consist of doing the project (direct implementation). An agency might actually install wind-powered generators at promising sites. It might give technical assis-

## Table 16.4. Tactics for Implementation

| | |
|---|---|
| Operations | Construction and installation |
| | Contracting construction and installation |
| | Technical assistance |
| | Research |
| Information/assistance | Planning and management |
| | Coordination |
| | Funding advice |
| Finance | Tax credits or penalties |
| | Tariffs or import restrictions |
| Standards | Establishing |
| | Publishing |
| | Monitoring and enforcement |

tance to people doing their own projects. It might do research itself or contract with another group to do research. Few government agencies actually have the people or facilities to carry out a large project in-house. The U.S. government contracts out almost all its projects. Many people believe costs are reduced if agencies actually do only small pilot projects and contract for the rest.

*Information/assistance* consists of making information, not only technical information, available to possible new users. City governments fostering home insulation give out much information. Part of this tactic is gathering data about existing projects. The data gathered include costs and names of vendors. Information can be disseminated by printed reports, by a telephone service, by seminars, and so forth. Social marketing, a topic of Chapter 17 fits here. A consultants bureau, such as an agricultural extension agency, might be established to meet more complex needs, with the consultants visiting sites and giving advice.

*Administration* is help in organizing and running projects. Some inventors, who handle technical matters well, do not have the talent or temperament to plan and manage projects. An agency could coordinate projects if several projects are working in the same area — avoiding duplication and encouraging cooperation. Finally, an agency could advise an inventor where to find funding if the agency did not have the funds itself. The Small Business Administration often gives management advice and assistance.

*Finance* includes giving tax credits for desired behavior. The Carter administration reduced taxes for people who installed energy-saving devices or insulation. Similarly, a government could tax heavily those things not desirable, as cigarettes are now taxed. In Third World countries, high tariffs are often used to discourage undesired imports. If high tariffs are not prohibitive enough, the import can simply be forbidden (which is really an administrative action, not a finance one). Not all tactics need to be punitive; the government of Bangladesh successfully encouraged the textile industry by reducing tariffs on power looms. Not only did the textile industry prosper as a result but also a whole new industry grew, originally to service the imported machines but eventually to manufacture replacements.

*Standards* is determining what is acceptable and customary in an industry. Planning the insulation of a house is easier, for example, if everyone agrees on what thicknesses of fiberglass will be commercially available. The first step in setting standards is getting representatives of the industry to agree on what will be standard. The second step is publishing the standards or using other means to make them known. The third step is monitoring and enforcement, checking that the products being sold meet standards. In some cases professional groups set standards but the government is often the most effective agency.

The particular set of tactics to be used depends on the situation — in particular, the funds available and the expertise and attitudes of potential users of the technology. Once tactics are chosen, their implementation must be scheduled, jobs assigned, and the rest of the required tasks performed.

Tactics analogous to those of the government agency previously described can be used by other organizations and will be needed for the same reasons. Even in companies, indirect implementation is often necessary if people are to become true believers of a new technology. A manager often achieves more by giving advice than by trying to do it herself/himself.

To implement an efficient village stove policy, the Malawi government might construct several stoves in various villages (an example of operations). The government could publish plans with costs (an example of information/assistance). It might send consultants to various parts of the country to give advice on how to use local materials (another example of information/assistance). It might tax imported stoves but reduce tariffs on the steel used to make stoves (finance). It might also set standards for design and construction so users know whether the stove they are building or purchasing will do what it should.

The same tactics would be useful in replacing diesel engines by small-scale wind or hydropower. Furthermore, financial tactics could be used — high tariffs for diesel fuel and engines and low tariffs for wind and water machines. If necessary some experimental work, part of operations, could be done — modifying imported designs to meet Malawian conditions.

# SUMMARY

Policy is made by simultaneously deciding on objectives, recognizing significant external factors, and recognizing significant internal factors. Alternative policies are developed next. An optimum policy is chosen from the alternatives. Finally, this optimum is put into practice.

When an optimum policy is being chosen, one estimates what the costs of each policy will be and what each policy will yield. One ensures that policies chosen will in fact yield enough. One also examines intangible factors, such as the effect on the culture of those involved and on the environment. One works out an organizational structure by deciding who will do what jobs, how good ideas will be communicated from project to project, and how problems will be recognized.

In the process of developing a policy, as many people as possible are involved to create confidence in the policy. When the policy is chosen it is widely promulgated within the organization so all members know the purposes and programs of the policy. The strengths of the chosen policy are made clear to people outside the organization, so they also will have confidence in it. These strengths, of course, are known because they were demonstrated when the policy was made.

When working with Appropriate Technology, it is especially important to give the people actually doing the work autonomy and support. Tight supervision reduces motivation and takes design responsibility from the person who knows the local situation best. Implementation of a policy may consist of actually building something or it may consist of supporting people who are doing the building. Government agencies nearly always do the latter. In either case, definite plans specifying who will do what and when are needed.

The process of making policy should accomplish several goals: The chosen policy should meet required material needs, an organizational structure should be created to implement the policy, and people should be convinced the chosen policy is a good one.

# REFERENCES

1. Chambers, R. (1997). *Whose Reality Counts: Putting the First Last*. Intermediate Technology Publications, London.

   *Discussion of why "development" projects have been unsuccessful, followed by a strategy for making future ones more successful by effectively involving the participants; some tough and lively writing.*

2. Drucker, P. (1964). *Managing for Results*. Harper & Row, New York.

*Drucker has written many sensible books about management. This one is especially pithy and emphasizes focusing on real needs.*

3. Forsyth, D. J. C. (1990). *Technology Policy for Small Developing Countries*. St. Martin's, New York.

   *The author is an economist and has served as a consultant in several Third World countries. The book reflects this background. It is especially useful because it describes specific tactics nations can use to implement a technology policy.*

4. Hazeltine, B. (1981, May 9). *Energy Planning Issues in Malawi*. Society for the Advancement of Science, Malawi.

5. Kuehn, T. J., and Porter, A. L. (Eds.) (1981). *Science, Technology and National Policy*. Cornell Univ. Press, Ithaca, NY.

   *A collection of articles describing various aspects of science policy and how it is formed. Not much help for planners but plenty for observers of Washington and other capitals. The list of tactics given in Table 4 is based on a list given by Eugene B. Skolnikoff in "The International Functional Implications of Future Technology," pp. 226–246.*

6. Mckenney, J. L., and Copeland, D. G. (1995). *Waves of Change: Business Evolution through Information Technology*. Harvard Business School Press, Boston.

7. Merkhofer, M. W. (1987). *Decision Science and Social Risk Management*. Reidel, Dordrecht.

   *Deals mostly with choosing alternatives; gives several mathematical approaches; the three examples are air pollution in the Los Angeles Basin, commercial cryptography, and choosing experiments for a space mission; critique of quantitative methods rather than an explanation of them.*

8. National Academy of Science (1976). *Systems analysis & operations research: A tool for policy & programs planning for developing countries,* Report No. PB 251 639 (prepared for Agency for International Development). Available from National Technical Information Service, Springfield, VA.

   *A general report on how mathematical methods can be used to help in planning; describes the approach but not the methods themselves; good place to start; examples are bus routing in an Indian city, water systems in the Indus basin, South Korean national economic models, agriculture in Mexico, and blood bank inventory control.*

9. Rich, D., *et al.* (1983). *The Solar Energy Transition*. Westview, Boulder, CO.

   *Papers on solar energy policy — how the federal government, the utilities, industry, and small business can contribute; insights into how government policy is formed are especially useful; argues for a Solar Lobby; uniformly high quality of papers; highly recommended.*

10. Rogers, E. (1983). *Diffusion of Innovation*, 3rd ed. Free Press, New York.

    *A standard work on technological change. The model given is more elaborate than Fig. 16.2. Rogers has had experience in Latin America, especially with population planning.*

11. Steiner, G. A., and Miner, J. B. (1977). *Management Policy & Strategy*. Macmillan, New York.

*A standard management reference on how to determine a policy for a company; clear and straightforward.*

# PROBLEMS

1. Develop a policy to reduce the harm done by automobile accidents in the United States. Give specific objectives, a plan, and the programs required.

2. Assume your objective is to get families in a rural county in the United States to raise half of their own food. List specific obstacles that objectors might raise to this policy.

3. Assume the objective of national policy for a Third World country is to become a fully industrialized nation. Two strategies are proposed. These are (much simplified) (i) to target a few industries and build state-of-the-art factories, which will employ and train a small percentage of the workforce; or (ii) to use Appropriate Technology to raise the general level of expertise for the whole country. Give advantages and disadvantages of both strategies.

4. Some economists believe private industry can do things better or cheaper than government. Does this apply to bringing Appropriate Technology to the Third World? Give arguments.

5. List and compare the managerial problems in opening, in a Third World country, a high-technology factory and an Appropriate Technology factory.

6. For what products, in the United States, is the allure of high technology relevant? Try to think of some products which are usually marketed on the basis of their high-technology features but for which high technology offers little functional value.

7. How would you answer a Third World leader who asks why her/his country should embrace Appropriate Technology if the United States does not?

8. How might Peter Drucker's observation, "It is better to do the right thing than to do things right," apply when a technological choice is being made?

9. How would you reply to someone who objected that policymaking is just a waste of time and that an organization such as a welfare agency or a nation does best by just waiting for opportunities to arise and then capitalizing on them?

10. List definite objectives for a specific nonprofit organization (e.g., a health clinic, university, church, or synagogue).

11. List external threats and opportunities for a specific nonprofit organization, perhaps the one selected in the previous problem.

12. Government agencies often have a reputation of treating each other as rivals — as protecting their turf. How could such rivalries be minimized? Can the rivalry serve a beneficial function?

13. Assume your objective is to improve health care in a rural county in the United States. What internal strengths and weaknesses of your organization would you be particularly concerned about?

14. Assume a team of five people is planning to implement a community garden project. The land has been identified but city approval is needed before gardening starts. Only people from local apartment buildings will be eligible to garden. It is now January 1. The intention is that the team will be engaged only for 1 year. Set up a schedule, a budget, and an organization.

15. If you were a Peace Corps supervisor in a Central American country, how would you interact with the volunteers, nearly all of whom are in rural areas working on different projects?

16. Outline the process of determining energy policy for a particular state in the United States. Start by choosing the state.

17. Assess the validity of the model of innovation given in Fig. 16.2 by comparing it with an actual popularization of an innovation, such as the use of computers at a college, passive solar water heating, or purchase of a VCR for home use.

18. If you were responsible for encouraging faculty members at a university to use new technologies, such as CD-ROMs, what would you do now? In what ways would your tactics change as time goes by?

19. Do you think it is common for a person at the working level of an organization to know her/his job better than the supervisor? Give examples in which such might be true. What is the function of the supervisor in these cases?

20. Assume you are producing a new kind of insulation that is cheaper and better than that now on the U.S. market. How would you let people know about it? Do the same means of publicity exist in the Third World?

21. Assume you want to insulate a house and would like the best insulation available. How would you learn about new materials? Do the same means exist in the Third World?

22. How would you decide what kind of car to buy if you were to buy one next year? Consider cost, size, appearance, and other factors. How would you make trade-offs?

23. Assume a member of a family or of a similar size group were to buy a car for the group. How would you ensure that the buyer's biases do not overly influence the decision?

24. Who might be the natural supporters of an Appropriate Technology constituency if one were to organize a lobbying group?

25. Pick several strategies from Table 16.3 that make sense in the United States and suggest how an agency of the U.S. government could implement them.

26. How would you apply the tactics of Table 16.4 to the problem of getting homeowners to recycle solid waste? State which tactics seem irrelevant.

# 17

..........

# THE PROSPECTS FOR APPROPRIATE TECHNOLOGY

.....................

**RURAL INDUSTRIES
INNOVATION CENTRE**

USE YOUR POWER WISELY

## THE FUTURE DIRECTION OF APPROPRIATE TECHNOLOGY

In this final chapter we take up three questions: Why is Appropriate Technology not taken more seriously? Why should we care if it is not commonly accepted? and What can be done to promulgate its use? These questions have surfaced earlier but are worthy of a more complete discussion.

# SHOULD APPROPRIATE TECHNOLOGY BE TAKEN SERIOUSLY?

We deal first with two common criticisms—that the motives of the proponents are frivolous or sinister. Could Appropriate Technology be simply a plaything for rich people? Marie Antoinette built a model of a farm in a corner of the gardens at Versailles where she and some of her court played at being sheep herders. Is Appropriate Technology a diversion for those who possess the benefits of high technology and do not have to worry about the dangers of a poor choice or could it be a strategy on the part of better off nations to keep themselves better off by ensuring that poorer countries do not become industrial rivals? Because these questions deal primarily with the motives of the proponents, they really miss the point. What is important are the results of the Appropriate Technology approach. Have small-scale, labor-intensive, locally controlled technologies improved the lives of the people involved?

Study of the results of Appropriate Technology is ambiguous. Although one can point to few conspicuous successes, one cannot point to many conspicuous failures. Much has been written about Appropriate Technology. After 20 years some successful projects are in existence but not as many as would be expected from the amount of writing or talking. In the days of vaudeville the question was asked about acts that seem good in rehearsal: "Will it play in Peoria?" It still is not clear if the Appropriate Technology approach will play in the global equivalent of Peoria.

## *The Colonial America Legacy*

Skeptics about the future of Appropriate Technology can point to what occurred in the United States. Did not the people, both Native Americans and Europeans, who settled North America originally practice Appropriate Technology? Was not Thomas Jefferson's vision a nation made up of self-reliant farmers using technologies they controlled? Is it not true that what amounted to Appropriate Technology existed in the United States approximately 200 years ago and has now nearly disappeared?

It is true that much of the hardware used by colonists in New England and the middle Atlantic States was similar to that described in the first part of this book. Waterwheels were built on nearly every stream. Handcrafted

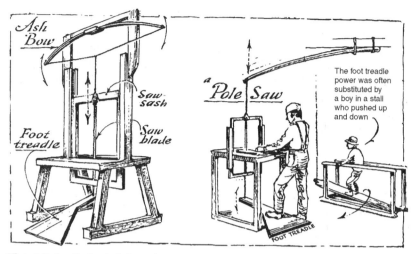

**Fig. 17.1. Colonial American tools. (Reproduced with permission of Ballantine Books [10].)**

tools were essentially the only ones available. Colonists were imaginative and skillful at making things (see Fig. 17.1) [10].

Settlers were especially adept at working with wood — carpentry, furniture making, ship building, and so forth. They were fortunate to have abundant forests and also knowledge of how to make axes, chisels, and knives. They were able to grow food for survival on land that is usually spurned by present-day farmers. Through much of New England forest one can find stone walls built by settlers. What is now forest was used then as fields or pasture. Colonial U.S. technology — hydropower, tools, agriculture, and others — does fit parts of our definition of appropriate. It was small scale, certainly labor-intensive, and built and maintained by the user.

Why did this way of living die out? First, much of it was, one might say, in an inappropriate situation. Many of the self-reliant farmers lived in New England or the hills of Appalachia where the soil was rocky and not ideal for farming. The previous occupants of the land had used it much more gently — both farming and hunting less intensively. The colonists, in many cases, wore out the land (i.e., drained the soil of nutrients), making continued farming even more difficult (an example of unsustainable development). Some of the best land used by settlers is in Pennsylvania where Amish people are still farming it, practicing to a large extent Appropriate Technology. Land in many other areas used by colonial settlers is currently lightly farmed.

Another reason colonial technology died out was that it could not meet all the needs of the colonists. Cash was required for textiles, iron pots and axes, and other necessities. This cash was earned by making things at home during the winter, for example, straw hats, shoes, potash, maple sugar, and nails (Fig. 17.2) [10]. Large-scale factories, however, which were established in the nineteenth century, could make most of these products cheaper and therefore eventually eliminated this source of cash. Farmers could not find other products that could be made competitively at home.

Part of the reason colonial technology disappeared was that people left the farms; an alternative way to gain cash was for someone in the family to work in a textile mill. These mills used cotton brought from the plantations in the South. The thread or cloth was sold nationally. Although the original mill designers had learned the business in England and copied English mills, soon Americans were able to design mills on their own. The mills attracted young people from the small farms to towns and cities. As we have seen in earlier chapters, this drain of young people from rural areas is currently taking place in the Third World. Were cash needs the only reason young people left the farms? A second need not met on colonial farms was social. It was difficult for people on isolated farms to get together, to learn new techniques from each other, and to find spouses. An oft-cited reason for the widespread and rapid acceptance of telephones in rural areas is how

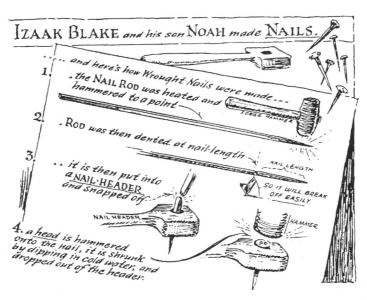

**Fig. 17.2. Making nails to earn cash. (Reproduced with permission of Ballantine Books [10].)**

it "brought farmers' wives in touch with each other as never before" [2]. An attraction for young women and men to leave home to work in a textile mill was an opportunity to meet new people and see another kind of life. A probable third reason young people left farms is that, just as in the Third World today, opportunities seemed more abundant in the cities, partly because new developments were occurring there.

Colonial technology has pretty much disappeared: What is its legacy? One legacy was the unleashed creative energy of the people who farmed and invented in New England. The experience of designing, building, and repairing farm machinery and household gadgets carried over directly to textile machinery. When that industry came to New England, both skill in making machinery and the resourcefulness in trying new devices were already present. The expression "Yankee Ingenuity," referring to this skill and resourcefulness, became known throughout much of the world. The same skill and attitudes are reflected in the successful high-technology industries now in New England. This heritage of invention, of course, went much further than New England. Henry Ford's Greenfield Museum in Michigan is stocked with colonial New England tools and machines, probably because Ford felt these were the foundation of the automobile industry which grew near the Great Lakes.

Colonial U.S. technology was successful, despite handicaps, in several ways. It provided sufficiently for the needs of the colonists so they survived and prospered. It led to the next stage in U.S. industrial history. It overcame handicaps, such as a terrain inhospitable to any sort of intensive farming. (The population density of Native Americans in New England was very small by present-day standards.) The accomplishments of the colonist are especially impressive when one considers how much more has been learned in the past 250 years of the science and basic technology they employed. The example of colonial U.S. technology does give confidence that Appropriate Technology can make lives better.

## The Threat from the Engineer

Another reason for wondering if Appropriate Technology has a future is exactly the Yankee ingenuity described previously—the desire of some engineers to elaborate a design so much that it is no longer Appropriate Technology. Functioning greenhouses can be made complex by sophisticated water-purification systems and computer-controlled heating and cooling. Some woodstoves for the Third World are currently made too complicated and expensive to be accepted by users. It is satisfying to design a complex thing and good engineers enjoy using state-of-the-art components. The result of an engineer's response to challenge can be an intricate design which is not small scale, controllable, or maintainable. Lewis Mumford [7], a social critic, uses an analogy to beavers. Beavers must gnaw or else their teeth grow too long and pierce the skull. Therefore, beavers cut down trees and build dams all the time. Engineers are the same way, says Mumford; they must be building, improving, or redesigning all the time. The problem for Appropriate Technology is that the redesign may be a better piece of engineering but meet real needs less well. Perhaps this urge to redesign is part of the allure of high technology described previously.

How can an Appropriate Technology device be saved from being overdesigned? Possibly by ensuring in the design process that the objectives of simplicity, control, and easy maintenance be explicit. An important aspect of the Appropriate Technology approach is that the design match the needs of the situation and that the design take into account factors other than engineering sophistication.

The basis of the three kinds of skepticism described—that Appropriate Technology proponents have other purposes, that Appropriate Technology was tried and eliminated in the United States, and that Appropriate Technology inevitably turns into inappropriate high technology—is the identification of Appropriate Technology with its hardware, that is, with the devices used. In truth, the real usefulness of Appropriate Technology may be the software (the approach)—the inclusion in the design process of broader factors such as the effect on culture and environment and the focus on the user's situation. One answer to skeptics is that they and others are paying too much attention to the gadgets and not enough to the approach itself. In other words, Appropriate Technology is more than the machines the skeptics point to. This answer amounts to stating that the reason for few conspicuous successes for Appropriate Technology is that it has not been fully tried. (The same observation has been made about most major religions.)

## WHY WE SHOULD CARE ABOUT THE FUTURE OF APPROPRIATE TECHNOLOGY

A reason we should care about Appropriate Technology is based on what we see in the future. Certainly, there are increasing numbers of people on the earth and thus greater threats on resources and the environment. Appropriate Technology can use less resources and be easier on the environment than the alternatives for the same output.

Appropriate Technology may also be beneficial for intangible reasons. Technological changes will certainly take place both in the United States and in the Third World. Better communication and transportation will mean people, ideas, music, and so forth will travel further and faster. The risk is loss of cultural diversity. Unless care is taken, communities all over the world will look the same, have the same music, grow the same food, and so forth. Another risk of better communications is the temptation to build larger organizations, to centralize more decisions so individual members of the organizations have less control. As technology becomes more complex, a threat develops that fewer people understand it. It is dangerous when only a few understand the essential industries of a society and the rest must proceed on trust. A complex technology is also a threat if it results in a split society, with a very few people having rewarding, challenging jobs and the great majority having meaningless, dull jobs. Appropriate Technology may be a way to make people's lives richer and safer.

A more general reason we should care about what happens to Appropriate Technology is that Appropriate Technology represents a choice. Appropriate Technology may not be the best choice in all situations but the first part of this book shows that cases do exist in which it is best. Because Appropriate Technology is often well matched to the needs of the Third World and the underdeveloped parts of the United States and Europe, e.g., blighted urban neighborhoods, it may offer a possibility of relative parity in standard of living without unrealistically large investment. We should care about the future of Appropriate Technology because the hardware promoted can be more effective than the alternatives, because it promotes valuable social attitudes, and because the approach reminds us to take a full range of factors into consideration when making a choice.

## Appropriate Technology and Social Marketing

We have seen that although Appropriate Technology has been successful in only a few places so far, the approach has value now and should have value as the world changes. What can a supporter do to ensure that it is considered when it might be useful? Marketing methods have been applied successfully to other social causes and may be helpful here.

To help clarify the marketing approach, two concrete examples will be considered. In both, some person or group will decide how to meet a need and we want that person or group to treat Appropriate Technology seriously before making a decision. One example involves the use of solar water heaters for rural homeowners. We think that these water heaters would produce sufficient hot water and that their use would benefit the region by reducing fuel oil consumption. The other example is encouraging the ministry of health in a Third World nation to promote the use of ventilated improved pit latrines (VIPs). In one case, a fairly large number of people (many homeowners) must be persuaded to use a new device. In the other we need to convince only a small group. (Of course, the ministry of health, if it does institute VIP latrines, will have a large education/persuasion campaign on their hands, but that is not our problem now.) How would marketing ideas help us in these two situations?

Marketing consists of two major efforts: (i) learning about the prospective user and (ii) deciding what to do in each of four areas — product, price, promotion, and distribution [5]. (It is easier to remember these areas if one replaces "distribution" by "place." One can then remember "4 Ps".)

Most of what will be described seems obvious. However, it is striking how often large parts of it are omitted. Engineers, especially, are prone to believe that a good idea will be accepted, even sought after, by potential users. Not true! The world in fact will not beat a path to the doorstep of the person who designs a better mousetrap. The best of ideas must be brought to possible consumers and presented convincingly. Maybe the reason Appropriate Technology has not played in Peoria is that nobody has tried to convince the people there.

The first part of marketing is focusing on prospective users. How does one learn about prospective users? Usually the best strategy is to talk to as many as possible. Find out real needs, find out conditions in which the technology will be used, and find out what the user can do herself or himself. The history of efficient woodstoves contains many examples of failure to do enough on-site research. Do the people using the stove perceive a need to conserve wood? Is the smoke of an open fire useful in repelling insects? Do mothers prefer a quick fire so as to have more time for something else or a slow fire which warms the room? Formal market research, a mail survey of rural homeowners, for example, may be useful if we are encouraging solar water heaters but it definitely needs to be supplemented by direct contact with the user. A former president of Campbell Soup spent as many Saturdays as possible in the local supermarket observing food purchasers, gaining greater insight on how his market judges his products.

If the potential user is an individual, such as the director of a ministry of health, it is just as important to find out her or his needs, attitudes, and concerns. What really is most important to the user? An example has been described from Tanzania of a manager who was generally committed to Appropriate Technology but believed himself to be under pressure to get results quickly from foreign assistance and so chose to implement a large hi-tech sugar refinery rather than small-scale sugar production [12]. These sorts of hidden pressures need to be recognized.

Political and cultural factors may be very important to decision makers and these factors may not be obvious. It would be naive to pretend that corruption, another sort of hidden pressure, does not exist among government officials, both in the United States and in the Third World. It may not be apparent how to deal with corruption but it is certainly worth knowing where it exists. An example of a political factor is a large investment in an automated factory made by the ruling party of a nation. Another political factor is that foreign assistance may pay a large part of a nation's bills and donor countries may not believe in Appropriate Technology. A third political factor might be that the system of land ownership permits only large farms to succeed because small landowners cannot muster political support. Some political or economic factors can be so dominant that the Appropriate Technology approach is unfruitful.

A final issue in learning about potential clients is to determine who actually makes the choice. Would the decision to push VIP latrines be made by the minister of health, by the president of the nation, or by a project director in the middle of an organizational chart? One needs to know so as to convince the right person or persons.

## Price, Product, Place, and Promotion

Once one understands the user, one puts together a program combining price, product, place, and promotion. For a nonprofit organization the pricing decision is usually straightforward — price is determined by cost. For a counterexample see Ref. [8], in which nonprofit organization promoting contraception elected to set its price for condoms equivalent to what consumers "paid for a cup of tea, a box of matches, or a cigarette." Over the long run, the cost of our approach must be shown to the user to be less than that of the alternative approach for the same output.

It goes without saying that the product must meet local needs — small-scale technology did not meet all the needs of colonial farmers and was superseded. The temptation for promoters of Appropriate Technology is to take a product which works well in a research station or in another part of the world and install it on a large scale, without local testing. Local testing and redesign though is absolutely necessary. VIP latrines, as mentioned previously, collected snakes in Kenya: an experimental model might have shown this. One risk of pushing an untested device on a user is that if it fails the user is likely to be suspicious of all future suggestions. It seems unconscionable to ask users to pay for experimental projects or for possibly ineffective products, as the people on the Marshall Islands were asked to pay for their photovoltaic systems.

Continual product improvement or adaptation is essential and it is best if the user is involved with the improvement. The product should be designed so the owner can modify it and adapt it to meet specific needs or use specific resources. Passive solar heating techniques, for example, are flexible enough to make specific rooms especially warm or well lit or to make use of existing large trees for shade. To summarize, product choices, in general, should begin with understanding the user, not with what the giver has available.

Place or distribution is concerned with how the product is physically moved to the user. It is also concerned with how much service goes with the product. If one wanted to have solar water heaters in rural areas would one sell through existing hardware stores, by mail, or send salespersons out door-to-door? Would one expect the owner to do the installation? Would local warehouses be essential? Is a network of repair specialists necessary or can existing local plumbers do maintenance if information is made available to them? Here, again, understanding of the user is essential. Especially for a new technology, frequent visits to consumers to see how the device is actually being used are important but are often neglected. Equally important is easy access by the user for assistance. Telephone numbers, a series of seminars, and a local office staffed by consultants are all reasonable ways of providing this assistance.

Promotion includes advertising. It may also include small gifts— buttons, shopping bags, calendars, and so forth. In Thailand, family planning was promoted in the villages in several ways: Traveling puppet shows were used and T-shirts with the program logo were distributed. Inches and centimeters were printed on pill packages which could then be used as rulers. This Thailand program is described in more detail in a page from a United Nations manual shown in Fig. 17.3 [13]. In the United States, TV, radio, or newspaper advertising could be used to promote solar heating.

A ministry of health would be approached differently by a promotional campaign, depending on the officials' reservations about the project. If credibility about the effectiveness of the hardware is a problem, then presentations with slides or videos of successful projects might be effective. Well-designed brochures might also be useful. If the minister of health doubts that the VIP latrines really improve health, then testimony from physicians or public health specialists, either at presentations or by letters, might be used. The choice of tactics, like all choices in this field, depends on the specific requirements of the situation. The planner must analyze systematically needs and then match tactics to those needs—which is exactly what marketing comprises.

## Rangpur–Dinajpur Rehabilitation Services

Some marketing ideas are illustrated by a project of the Rangpur–Dinajpur Rehabilitation Services (RDRS) in Bangladesh (described in Chapter 6). RDRS is a branch of an international nonprofit organization. RDRS people

*Box 38  Family planning in Thailand – a social marketing approach*

In the early 1970s, family planning in Thailand was restricted to servicing through medical channels and mention of the subject tended to cause social embarrassment. Today, public discussion of, and advertisements for, family planning methods have become commonplace. This change has largely resulted from the activities of a nongovernmental organization, the Population and Community Development Association (PDA), which has shown impressive achievements in promoting social awareness of family planning and the widespread adoption of contraception.

The Association is a non-profit and largely voluntary organization with over 300 staff members and 16 000 village volunteers working from five offices.

A new understanding of social motivation, combined with a large measure of humour and audacity, has led to the development of a wide range of culturally adapted social marketing strategies, implemented through the PDA programmes. Between 1974 and 1981, one of the programmes, the Community-Based Family Planning Services (CBFPS), grew from a pilot undertaking into a network of local self-help schemes reaching over 16 200 villages in 158 districts with a population of 17 million. Its consumer-oriented approach has been adopted within the National Family Planning Programme for extending coverage and maintaining high user rates.

Since the inception of the programme, the main message has linked population growth to low standards of living, on the one hand, and family planning to economic advantages on the other.

This message is conveyed by creative use of virtually all possible communication channels, leading to a high level of public awareness of family planning and of the PDA programmes. Face-to-face education is undertaken by the village distributors, using the information, and motivational and publicity materials with which they are provided.

In addition, television and radio broadcasts are made on issues related to family planning, and many programmes of general interest close with reminders about using contraceptives. In school, children learn about family planning and may be taught, for example, a song describing the hardships resulting from having too many children.

Troupes of traditional itinerant entertainers perform puppet plays containing family planning messages in villages all over the country; T-shirts and other promotional materials carrying family planning messages have been distributed at formal state dinners and sent to foreign heads of state, emphasizing the legitimacy of the programme and the support of the government.

To convey the message through the various communication channels, it was necessary to overcome the taboos surrounding birth control techniques and the social embarrassment at discussing them openly. In Thai culture, humour and joy were found to be the best means. Carnivals, games, raffles, village fairs and weddings serve as occasions to promote family planning joyfully. Inches and centimetres are printed on contraceptive pill packets so that they can be re-used as rulers. Sheets, pillow-cases, piggy banks and business cards are all printed with family-planning catchwords.

*Source: Dhul, L.J.* Social communication, organization, and community development — family planning in Thailand. *Assignment children, No. 65/68, 1984.*

**Fig. 17.3. Page from United Nations manual. (Reproduced with permission from UNICEF [13].)**

had spent much time in rural Bangladesh and understood the situation of the farmers (the product met local needs). The agency subcontracted to have the pump built at a low cost in Dhaka, the capital, so the tube wells were within the price range of many of the region's farmers.

The marketing strategy used by RDRS in introducing these pumps was to demonstrate their use at various locations in the districts, as in Fig. 17.4. Pumps were set up near bazaars where many people could see their effectiveness and also in front of the houses of the influential people of the villages to show that these pumps were endorsed by them as well. The agency was prepared to offer loans to farmers to purchase the pumps but such was not necessary because the price was sufficiently low. In addition, RDRS provided technical assistance for the setup and maintenance of these pumps.

The marketing campaign worked. Within 1 year from the introduction of the bamboo pumps, they were being widely used by farmers in the two northern districts. Furthermore, entrepreneurs in Rangpur and Dinajpur began the business of manufacturing the steel components of the pumps. Thus, it was no longer necessary for RDRS to provide these components from Dhaka; the farmers themselves set up new pumps and maintained them. After the introduction of these pumps, the area irrigated in the two districts also increased, as did crop yields for the small farmers. Also, employment increased because of the manufacture of the steel components.

**Fig. 17.4. Demonstration model of RDRS bamboo tube well.**

This project illustrates some of the ideas of social marketing. The product was chosen to meet a real need of the users — water for irrigation. It was tested and demonstrated in the local area. The price was low enough for the users. It was promoted at places where local users attend. Technical assistance was made available in the regions where the farmers lived.

## WHAT WILL HAPPEN TO APPROPRIATE TECHNOLOGY?

The authors are optimistic about the future of Appropriate Technology both because the approach has had some successes and because it is becoming generally regarded as a worthwhile possibility. One hardware success is housing in the United States. In the past decade many home-owners have insulated and plugged cold air leaks, improving comfort through conservation rather than through more elaborate heating systems. It is now generally expected that new buildings will be energy efficient. In much of the Third World, oral rehydration therapy is common. Many of the small-scale agricultural machines developed at IRRI have been well accepted. Village factories in China have contributed to that nation's development. A UNICEF report [13] describes successful projects dealing with energy, water supply, nutrition, health, basic education, and cash earnings. The rationale for the approach has also proven successful. Organizational designers in the United States are often aim at small, autonomous groups rather than large centralized organizations. Technology planners are considering, much more thoroughly than in the past, the interaction between the technology and the people involved. Articles mentioning Appropriate Technology seem to appear more often than they did — the concept is generally known and accepted.

The other reason for optimism is that people now seem to be doing Appropriate Technology better. Recent hardware designs are simple and elegant rather than complicated, as they sometimes used to be. More attention is paid to actual needs. Designers are more realistic about which needs Appropriate Technology can meet. Specialists seem to understand better how to design a program implementing a technology change, for example, what kinds of support users need or how to accomplish technology transfer. Also, more is understood about which economical and political environments will be most hospitable to Appropriate Technology. Appropriate Technology, of course, is not the optimum choice in every situation. People are learning to recognize both the technology factors and the social, economic, and political factors that make it a good choice.

# REFERENCES

1. Cronon, W. (1983). *Changes in the Land, Indians, Colonists, and the Ecology of New England.* Hill & Wang, New York.

   *Compares the Native American's treatment of the environment with the colonists; serious history; well written and absorbing, with many lessons.*

2. Fischer, C. S. (1992). *America Calling: A Social History of the Telephone to 1940.* Univ. of California Press, Berkeley.

3. Fox, K. F. A., and Kotler, P. (1980, Fall). The marketing of social causes: The first 10 years. *J. Marketing* **44**, 24–33.

   *Describes briefly, with examples, how social causes are marketed; aimed really at marketing people but a useful overview for anyone; addresses other issues such as major criticisms of social-cause marketing.*

4. Hazeltine, B. (1993, September). Small scale technology in Botswana: Utility and support, SAMS Working Paper Series. School of Accounting and Management Studies, University of Botswana, Private Bag 0022, Gaborone, Botswana.

5. Kotler, P. (1976). *Marketing Management.* Prentice Hall, Englewood Cliffs, NJ.

   *A basic and thorough book on marketing; should certainly be read before starting a campaign or doing anything serious with marketing.*

6. Macauley, D. (1983). *Mill.* Houghton Mifflin, Boston.

   *In magnificent pictures, a succession of textile mills on the same site are described. All but the last are hydropowered. One learns the technology and the social history of the region (southern New England). One gains a clear sense of how change occurred: This is a book nearly everyone will enjoy and learn from.*

7. Mumford, L. (1963). *The Highway and the City.* Harcourt, Brace & World, New York.

   *Mumford is a well-known critic of modern technology, as well as architecture, who has written several books. This book, a collection of magazine articles, gives the flavor of his thought and is a good place to start.*

8. Rangan, V. K. (1985). Population services international, Case no. 9-586-013. Harvard Business School, Boston, MA.

9. Raup, H. M., and Carlson, R. E. (1941). *The history of land use in the Harvard forest.* Harvard Forest Bulletin No. 20. Harvest Forest, Petersham MA.

   *Describes how a colonial New England village grew and declined from 1733 to 1907 by examining chiefly deeds to the land. The population was greatest in 1840 (1775 person) and went from 707 in 1765 to 757 in 1910; bare bones on cultural life but much can be inferred.*

10. Sloane, E. (1965). *Diary of an Early American Boy.* Ballantine, New York.

    *Sloane draws wonderful pictures (see Figs. 17.1 and 17.2). He also writes good history. The basis of this book is a diary written in 1805 which was found in a Connecticut attic. Sloan has written other books, some about tools. This one is the most fun, but all are recommended.*

11. Stewart, F. (1987, Summer). The case for Appropriate Technology: A reply to R. S. Eckaus. *Issues Sci. Technol.*, 101–109.

*Stewart is a respected economist interested in the Third World. The article mentions some of the successes of Appropriate Technology and gives a rationale for its continued support. Parts of this chapter are based on the article.*

12. Stewart, F. (Ed.) (1987). *Macro-Policies for Appropriate Technology in Developing Countries*. Westview, Boulder, CO.

*Eight case studies of the influence of political and economic policy on the success of Appropriate Technology, plus an introduction and conclusion. The example of the Tanzanian manager in this chapter comes from a comparison of Kenya and Tanzania, which had different industrial policies but ended up doing similar things because of pressures on local managers. This book is clear and worthwhile.*

13. UNICEF (1983). *Appropriate Technology for Child Survival*. United Nations, New York.

*Report of a workshop held in Nepal; describes what has worked and analyzes why; focus on software rather than hardware; valuable particularly because it tells what has really happened; examples includes stoves, forestry, and biogas.*

14. Dhul, L. J. (1984). Social communication, organization, and community development — Family planning in Thailand. In *Assignment Children*, No. 65/68. UNICEF, Villa Le Bocage, Palais des Nations, 1211 Geneve 10, Suisse.

# PROBLEMS

1. The accusation has been made that the environmental movement is basically a hobby for the well-off, who have little to lose if industrial growth is slowed. Based on your personal experience, is the accusation true? If it were, what should a concerned person do?

2. In what ways does the United States gain or lose if a Third World country successfully implements Appropriate Technology?

3. Were the Native Americans in New England, just before the colonists arrived, using an Appropriate Technology? Compare their use of natural resources with those of the colonists. On the basis of the comparison what can be said about the long-term effectiveness of the Native Americans' and the colonists' technologies?

4. Explain how the farming practices of the Amish people at the present time relate to Appropriate Technology.

5. In what ways were water-powered textile mills in colonial New England consistent with our definition of Appropriate Technology? What would have to be true to classify them as being an Appropriate Technology?

6. On the basis of U.S. history can it be argued that Appropriate Technology is a good foundation for a nation that eventually wants to be heavily industrialized?

7. If you were a present-day regional planner working for a region composed of small

distinct farms, how might you meet the social needs of the residents?

8. What influences may exist which encourage an engineer to use an elaborate rather than a simple design when both are equally effective?

9. Describe situations in which engineers cause harm by continually designing, building, and improving.

10. Suggest some difficulties that may arise if current trends in technological change continue. How can Appropriate Technology ameliorate these difficulties?

11. Comment on whether an ethical issue exists in using marketing techniques to "sell" social causes such as Appropriate Technology.

12. What kind of market research, in the broad sense, would you do if you wanted methane digesters to be used on farms in an African country?

13. If you were assigned the job of installing a photovoltaic system on the Marshall Islands, what would you want to find out about users' needs?

14. If a person wanted to sell a solar water heater to the household in which you live, what information should that person seek? (You might start by stating who in the household actually makes the buying decision.)

15. Many radios bear the notice "No User Serviceable Parts Inside." Is such a design strategy sensible for products sent to the Third World? Explain.

16. Develop a plan for distributing and servicing solar water heaters in rural areas of the United States. First list your assumptions about the needs and attitudes of the potential consumers.

17. Develop a plan for promoting the use of vegetable gardens in an urban area. First list your assumptions about the needs and attitudes of potential clients.

18. Pick a particular Appropriate Technology (e.g., solar energy). Forecast how the technology will change or how its use will change if current trends continue. Decide what changes you would like. Devise a strategy so what you would like to happen actually happens.

19. List two or three things you would want to see happen in the world, outside your immediate life. Is Appropriate Technology consistent with these wishes? Helpful? Inconsistent? Harmful? Irrelevant? Explain.

# INDEX